Site-directed mutagenesis as a tool for functional characterization of the Ammonium Transport Protein Amt-1 from *Archaeoglobus fulgidus*

Dissertation

von Daniel Čebo

aus Belgrad

angefertigt

am Institut für Organische Chemie und Biochemie

der Albert-Ludwigs-Universität Freiburg i. Br.

Freiburg 2010

1

Referent	Dr. Susana Andrade
Korreferent	Prof. Dr. Oliver Einsle
Tag der mündlichen Prüfung	15.09.2010

Table of contents

1. List of Abbreviations

Semantics

%	percentage
*g	multiple of gravitational acceleration
°	degree
°C	degree Centigrade
bp	base pair
A	ampere
Å	Angstrom (1 Å = 10^{-10} m)
A_u	absorption unit
Da	Dalton
g	gram
h	hour
L	liter
M	molarity
m	meter
min	minute
psi	pounds per square inch (1 psi ≈ 0.07 bar)
sec	second
V	volt

Amino acids

A	alanine	M	methionine
C	cysteine	N	asparagine
D	aspartate	P	proline
E	glutamate	Q	glutamine
F	phenylalanine	R	arginine
G	glycine	S	serine
H	histidine	T	threonine
I	isoleucine	V	valine
K	lysine	W	tryptophane
L	leucine	Y	tyrosine

General abbreviations

A. fulgidus	*Archaeoglobus fulgidus*
Amt	ammonium transport protein
ADP	adenosine diphosphate
ATP	adenosine triphosphate
BCA	bicinchoninic acid
BCIP	5-Bromo-4-chloro-3-indolyl phosphate
CMC	critical micelle concentration
ddH_2O	double deionized water
$D_{12}DM$	*n*-dodecyl-β-D-maltopyranoside
DNA	deoxyribonucleic acid
E. coli	*Escherichia coli*
EDTA	ethylene-diamine-tetra-acetic acid
GS/GOGAT	glutamine sythetase/glutamate synthase system
HCl	hydrogen chloride
IPTG	isopropyl β-D-1-thiogalactopyranoside
LDAO	lauryl dimethylamine n-oxide
LB	Lysogeny Broth
MA	methylammonium
NaCl	sodium chloride
NH_3	ammonia
NH_4^+	ammonium
NPT	nitrotetrazolium blue
NTA	nitroloacetic acid
OD_{600nm}	optical density at a wavelength of 600 nm
PEG	polyethylene glycol
PVDF	polyvinylidendifluoride
RNA	ribonucleic acid
SDS	sodium dodecyl sulfate
SDS-PAGE	sodium dodecyl sulfate polyacrylamide gel electrophoresis
TEMED	N,N,N`,N`-Tetramethylethylenediamine
Tris	tris(hydroxymethyl)aminomethane

Triton X-100	polyethylene glycol p-(1,1,3,3-tetramethylbutyl)-phenyl ether
UV	ultraviolet
v/v	volume per volume
w/v	weight per volume

2. Zusammenfassung

Ammoniumtransportproteine (Amt-Proteine) bilden eine hochkonservierte Familie von integralen Membranproteinen, die in allen Domänen des Lebens zu finden sind. Funktionelle Studien mit Mitgliedern dieser Familie führten zu kontroversen Ergebnissen sowohl die chemische Zuordnung der von ihnen transportierten Spezies (NH_4 oder NH_3), als auch der Klassifizierung dieser Proteine als Kanäle oder sekundäre aktive Transporter betreffend.Bisher wurden vier Amt-Strukturen mit hoher Auflösung gelöst: AmtB des Eubakteriums *Escherichia coli* [1, 2], Amt-1 aus dem hyperthermophilen Archaeon *Archaeoglobus fulgidus* [3], the Rh50 aus *Nitrosomonas europea* [4, 5], und das humane Rhesus Glykoprotein RhCG [6].

Alle Proteine weisen eine bemerkenswert hohe strukturelle Homologie mit 11 transmembranen Helizes pro Monomer auf. Eine schmale, nicht durchgängige und hauptsächlich hydrophobe Pore lokalisiert im Zentrum eines jeden Monomers des trimeren Amt-1, wird als Substratkanal vermutet. Die Selektierung und Rekrutierung des Substrats NH_4 oder NH_3 findet wahrscheinlich am extrazellulären Kanalausgang aufgrund einer Kation-π wechselwirkung statt. Diese wird durch Wasserstoffbrücken zwischen den Seitenketten von W137 und S208 stabilisiert. Der Weitere Fortlauf des Kanals wird durch F96 and F204 blockiert. Im Inneren der hydrophoben Pore in Richtung des Cytoplasmas werden die Seitenketten von zwei hochkonservierten an den Hohlraum angrenzenden Histidinresten (H157 and H305) durch Wasserstoffbrücken miteinander verbunden. Die Rolle dieser Histidinreste ist bisher noch unbekannt.

Um den Transportmechanismus der Amt-Proteine besser zu verstehen, wurde Af-Amt-1 als Model verwendet. Durch Kombination von zielgerichteter Mutagenese mit biochemischen, biophysikalischen und kristallographischen Methoden sollte die Validierung eines solchen Rekrutierungsortes selbst, die Möglichkeit von Ammoniumdeprotonierung- und Reprotonierungsvorgängen als auch die ionische Form des Substrats sowie die Rolle des cytosolischen C-Terminus beim Transport und dessen Funktion in der Interaktion mit regulatorischen Proteinen wie GlnK untersucht werden.

Summary

Ammonium transport proteins (Amt) form highly conserved family of integral membrane proteins present in all domains of life. Functional studies with members of this family have yielded controversial results with respect to the chemical identity of the transported species (NH_4 or NH_3) as well as to the classification of the proteins as channels or secondary active transporters. Presently, four Amt structures were solved at high resolution: the AmtB from the eubacteria *Escherichia coli* [1, 2], the Amt-1 from the hyperthermophilic archaeon *Archaeoglobus fulgidus* [3], the Rh50 from *Nitrosomonas europea* [4, 5], and the human Rhesus glycoprotein RhCG [6].

All proteins share a remarkable degree of structural homology with 11 transmembrane helices per monomer. A narrow, non-continuous and mainly hydrophobic pore, located at the center of each monomer of the trimeric Amt-1, is thought to be the substrate channel. At the extracellular channel cleft, ammonium selectivity and recruitment presumably occurs via a cation-π interaction with the side chain of W137 stabilized by hydrogen bonds to the side chain of S208. Below, F96 and F204 block the channel continuity. Further into the hydrophobic pore leading to the cytoplasm, the side chains of two highly conserved histidine residues (H157 and H305) are bridged by an H-bond and lie adjacent, with their edges pointing into the cavity. The role of these histidine residues is not yet understood.

To understand the transport mechanism of Amt proteins in general, we are using the *Af*-Amt-1 as a model. By combining site-directed mutagenesis and biochemical, biophysical and crystallographic methods we are investigating the validation of the recruitment site as such, the possibility of ammonium deprotonation and reprotonation events, the substrate ionic form or the role of the C-terminus in transport as well as its function in the interaction with regulatory proteins such as GlnK.

3. Introduction

3.1. Identification of Ammonium Transport proteins

Various organisms can use many nitrogen sources, including ammonia, nitrate, dinitrogen, and a variety of amino acids and other nitrogenous organic compounds. While animals only use nitrogen containing amino acids as nitrogen source, microorganisms and plants are in the most cases depending on the uptake of ammonium. One exception is the reductive dinitrogen fixation. This complex process is carried out by nitrogen-fixing bacteria present in the soil which are able to reduce N_2 to NH_3 using the nitrogenase complex system. This process requires the hydrolysis of 16 equivalents of ATP and is accompanied by the co-formation of one molecule of H_2 [7]:

$$N_2 + 8\,H^+ + 8\,e^- + 16\,ATP \rightarrow 2\,NH_3 + H_2 + 16\,ADP + 16\,P_i$$

Ammonium can be generated from molecular nitrogen by nitrogen-fixing bacteria in some plant cells as well, such as rhizobia in legume root nodule cells, and ammonium transporter proteins are implicated in the transfer of ammonium from the bacteria to the plant cytoplasm [8].

The transport of ammonium across cell membranes is a process of fundamental importance in almost all living organisms. Because the charged NH_4^+ and uncharged NH_3 molecules are in equilibrium with a pK_a of 9.25 in aqueous solutions, NH_4^+ is the predominant form (c.a. 99 %) at a typical cytosolic pH (Figure 3.1). At high ammonium concentrations in the environment (for example, above 1 mM), passive membrane permeation of the uncharged species can be sufficient to promote cell growth in the cases of enteric bacteria and fungi. However, at lower concentrations, when diffusion becomes insufficient for sustaining cell growth, ammonium transport proteins (Amt) are needed [9].

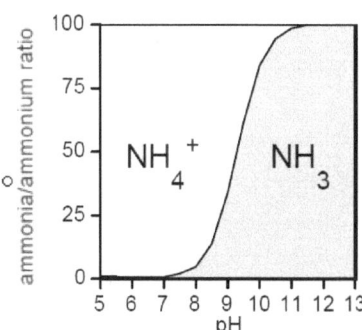

Figure 3.1: Effect of pH on the ammonium/ammonia equilibrium. Ammonia, the uncharged form is relatively lipid soluble and, in general, diffuses freely across cell membranes. At a physiological pH of 7.0, ammonium, the charged form is predominant although 1-2 % of total is in the uncharged form. At a pH of 9.25, both charged and uncharged forms are in equilibrium. Picture taken from reference [10].

The first ammonium activity was discovered in filamentous fungi [11]. The uptake activity was detected only under starvation conditions and was reported by the substrate analogue ^{14}C-methylammonium that was then shown to accumulate against its concentration gradient.

In 1994, the first two sequenced *amt* (ammonium transporter) and *mep* (methylammonium/ammonium permeases) genes were sequenced from *Arabidopsis thaliana* and *Saccharomyces cerevisiae*, respectively [12, 13, 14]. Eight years later, Merrick and co-workers overexpressed and purified AmtB by Ni^{2+} affinity chromatography from *E. coli* cells in large amounts [15]. They showed that AmtB forms highly stable trimers in the membrane that remain preserved during the purification procedure. The crystal structure of AmtB was later solved by two independent groups [1, 2]. Currently, three other Amt structures are known at high resolution: the Amt-1 from the hyperthermophilic archaeon *Archaeoglobus fulgidus* [3], the Rh50 from *Nitrosomonas europea* [4, 5], and the human Rhesus glycoprotein RhCG [6].

With increasing knowledge accumulated on these proteins and with further genome sequences becoming available, it became clear that Amt proteins are ubiquitous and form an independent family of membrane proteins [16]. For example, it was a study based on sequence analysis that revealed that the human Rhesus (Rh) blood groups of proteins are related to the Amt proteins [17]. Functional complementation experiments performed later demonstrated that human Rh proteins can indeed rescue the growth of a yeast mutant deficient in ammonium uptake [18] and thus function as Amt proteins The Rh family [19] comprises the erythroid blood group antigens and non-erythroid homologues that are expressed in the kidney, liver, skin, ovary, testes, and central nervous system [20, 21]. Although most evidences suggests that Rh proteins are part of the, often called, Amt/Mep/Rh family, they also have been proposed by some researchers to be gas channels for CO_2 [22-25]. Supportive, but controversial, evidence for this is that at elevated CO_2 concentration the expression level of Rh1 in *Chlamydomonas reinhardtii* increases [26]. Since RhAG homologues are also found in slime mode, sponge, nematode and fruit fly, these proteins probably derived from Mep/Amt proteins and originated later in evolution from a RhAG-like ancestor [27, 28] (Figure 3.2).

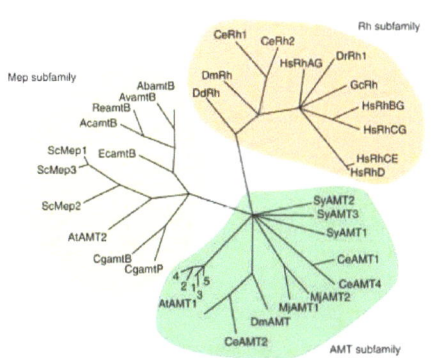

Figure 3.2: Phylogenetic tree of selected members of the Amt/Mep/Rh family, the family of ammonium transport proteins and related proteins. Figure taken from [29].

3.2. The Ammonium Transport Proteins of *Archaeoglobus fulgidus*

A. fulgidus is a sulfur-metabolizing anaerobe that grows optimally around 83 °C. This organism can produce biofilms when subjected to environmental stresses such as extreme pH or temperature, high concentrations of metals, or the presence of antibiotics, xenobiotics, or oxygen [30]. High-temperature sulfate reduction by *Archaeoglobus* species contributes to deep subsurface oil-well 'souring' by iron sulfide, which causes corrosion of iron and steel in oil-and gas-processing systems.

In 1997, the whole genome of *A. fulgidus* was sequenced [31] and the analysis of its 16S and 23S rRNA sequences showed a close relationship to methanogenic *Archaea* [32]. The *A. fulgidus* genome is a circular chromosome roughly half size of that from *E. coli* with 2.18 Million base pair (bp) long and containing approximately 2486 predicted genes. One observation about its genome analysis is the absence of several important eubacterial biochemical pathways such as the glycolysis, the pentose phosphate, and the gluconeogenesis and citric acid pathways. Furthermore, there is neither energy conserving pathways of the nitrogen cycle nor a nitrogenase system. In addition, three homologs of Amt proteins are present. Several other organisms like *S. cerevisiae* [13, 14], *Lycopersicon esculentum* [33] or *Methanococcus acetivorans* [34] have multiple *amt* genes within their genome. The reasons for this *amt* gene multiplicity are not yet demonstrated. It is supposed that the Amt proteins differ in their affinity to the substrate, such that the cell is able to control the ammonium uptake into the cells, as a response to environmental changes [35].

In the genome of *A. fulgidus* the three ammonium transporter genes are found in two separate loci *(amt-1, amt-2* and *amt-3)*, each one followed by a gene that belongs to a P_{II} family member *(glnK)* (Figure 3.3). Among the three GlnK proteins from *A. fulgidus*, GlnK1 and GlnK3 share 98% amino acid sequence identity while GlnK2 is the most distinct (60% with GlnK1 and 59% with GlnK3). The genomic location of GlnK2 in a gene cluster with the ammonium transporter Amt-2, a putative zinc hydrolase (AF1748), and the ammonium transporter Amt-3 followed by its corresponding GlnK-3 protein suggests that GlnK and Amt proteins are specific interaction partners, as has been shown for the GlnK/AmtB pair [36].

Figure 3.3: Organization of the *glnK-amt* operon in *Archaeoglobus fulgidus*.

Amt-1 and GlnK-1 are encoded by the AF0977 and AF0978 genes respectivelly, whereas the two other transporters are located in a different region and are encoded by AF1746/AF1747 (Amt-2, GlnK-2) and AF1749/AF1750 (Amt-3, GlnK-3). The gene AF1748 lies in between the two gene pairs of AF1746/AF1747 and AF1749/AF1750, probably encoding for a zinc containing hydrolase. Figure taken from reference [3].

The amino acid sequences of Amt-1 and Amt-3 show 64.2 % sequence identity, while Amt-1 and Amt-2 (39.7 %), or Amt-2 and Amt-3 (40.6 %). Amt-1 and Amt-3 have eleven transmembrane helices without a leader peptide cleavage side. Amt-2 has a 19 amino acids longer cytosolic C-terminus compared to Amt-1 and Amt-3. It has been proposed that this region has regulatory functions [37]. Amt-2 has higher homologies (42.4%) to the *E. coli* Amt-B (than the two others) that also shows a longer C-terminal region. (Figure 3.4).

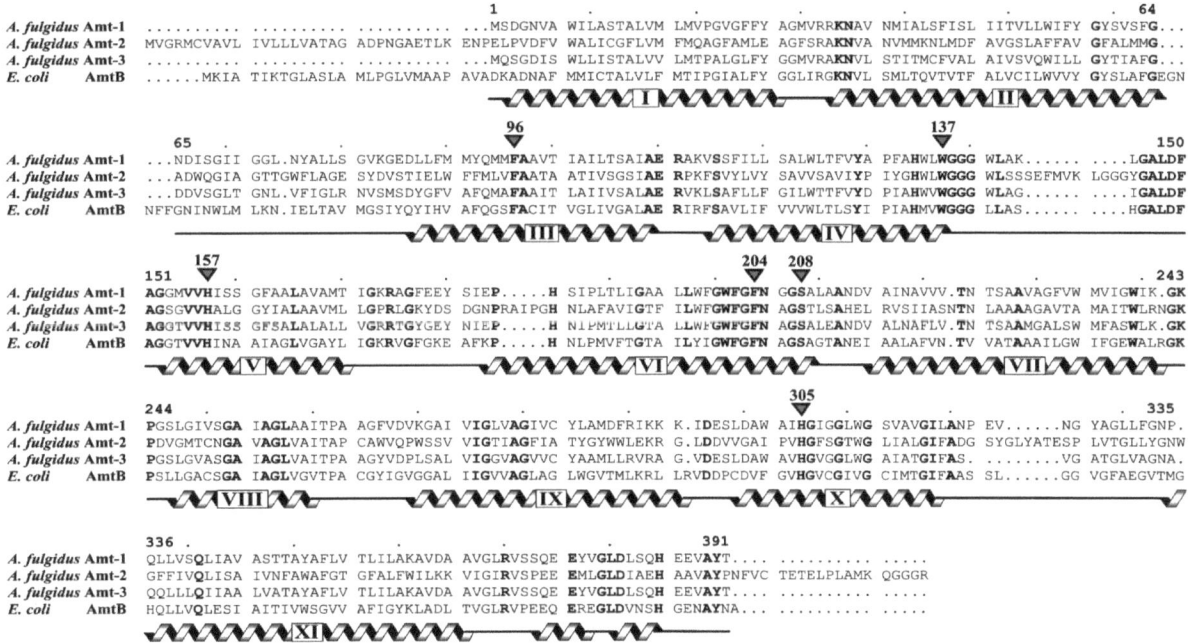

Figure 3.4: Sequence alignment of the ammonium transport proteins Amt 1, -2, -3 of *A. fulgidus* and AmtB of *E. coli*.

Highly conserved amino acids are highlighted in black and the position of Amt-1 amino acids that are predicted to have functional relevance is shown by an arrow. The outlined helices correspond to the data obtained from the crystal structure of Amt-1. Figure taken from reference [3].

3.3. Crystal structure of the Amt-1 from *Archaeoglobus fulgidus*

The X-ray structure of Amt-1 from *A. fulgidus* was initially determined at 1.54 Å [3], and later improved to 1.45 Å. Crystal structures with and without ammonium, methylammonium were also solved by soaking these molecules into the native protein crystals [3]. Crystal structure with a hydrophobic gas, xenon, has been solved, as well [3].

The tertiary structure shows that the protein crosses the membrane with eleven helices and that it homotrimerizes in the quaternary structure. Each monomer incorporates a hydrophobic channel in between relatively polar cytoplasmic and periplasmic vestibules (Figure 3.5).

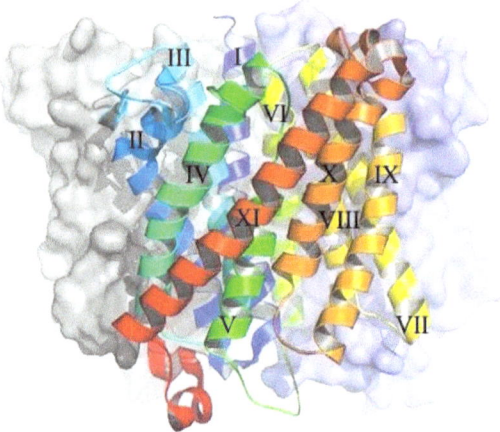

Figure 3.5: **Overview of *A. fulgidus* Amt-1 protein structure.** The picture shows the trimer of *A. fulgidus* Amt-1 protein with two monomers in surface representation and the monomer in the front in cartoon representation. The cytosolic part is below and the extracellular part is above. The different helices of the protein chain are marked with different colours from the N-terminus (in blue) to the C-terminus (in red). Helices are numbered from I-XI.

Consistent with the "positive inside-out" rule for membrane proteins [38] the trimer of Amt-1 has a net negative charge on the periplasmic surface and a net positive charge of on the cytoplasmic side. The residues from helices VI, VII, VIII and IX of one monomer interact with the helices I, II and III of neighbouring monomers, so that the protein homotrimerizes in a threefold symmetric fashion. The protein longest, eleventh helix with a length of more than 48 Å, stabilises the protein by crossing the two halves on the outer surface of the trimer. The structure also revealed a pseudo-twofold symmetric organization within the monomer itself, in which membrane-spanning helices I-V are related to helices VI-X by a quasi-twofold axis that lies in the mid-membrane plane. This pseudo-twofold architecture is well known from other membrane proteins like the lactose permease [39] or the ClC chloride channels [40], among many others.

Although there is no open channel crossing the protein (Figure 3.6) a closer inspection of the structure strongly indicated that substrate translocation occurs through a pore located at the center of each monomer right between the pseudosymmetry-related helical domains. The pore shows a vestibule on its extracellular and intracellular sides. The extracellular vestibule is supposed to contain the substrate recruitment and selectivity side.

The recruitment and selectivity site (Figure 3.6) is supposably formed by the aromatic ring of W137 and S208 [3]. A cation π-interaction between W137 and NH_4^+ stabilised by a hydrogen bond to S208 could explain the high affinity to NH_4^+ and the exclusion of other monovalent cations, such as K^+ (no hydrogen bonding interactions) and water (not a cation). Direct experimental evidence for this binding site has not yet been obtained and an assignment based on crystallographic data is impossible because water and ammonium have an identical numbers of electrons.

Further bellow, the transport pathway is blocked by the side chains of two conserved residues F96 and F204 (Figure 3.6). F204 has elevated B factor in comparison to the surrounding amino acids, indicating an increased mobility of this residue in the structure [3]. Based on this observation, it is supposed that some degree or conformational rearrangements has to occur, so that the substrate can pass this barrier and reach the cytoplasmic side [3].

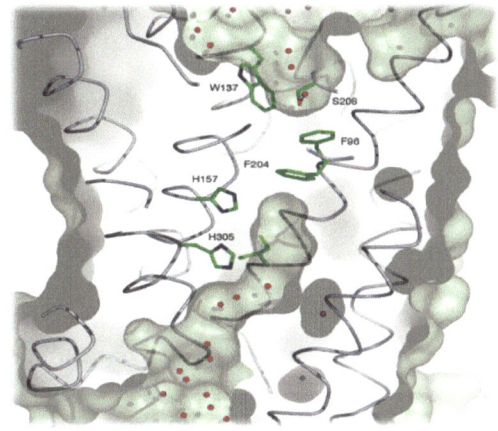
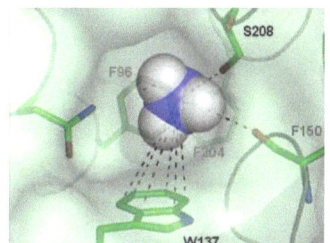

Figure 3.6: Inside view of one monomer channel _A. fulgidus_ Amt-1 protein.

Left: The outer (up) recruitment site, and inner cytoplasmic (down) vestibules are highlighted by a hypothetical cut, half way through the protein surface (green). NH_4^+ presumably forms cation π-nteraction with W137 stabilized by hydrogen bond to the side chain of S208 on the extracellular side of Amt-1 (at the top). Substrate is blocked from passing through the channel by the side chains of F96 and F204, but once this obstacle is passed a channel leads directly to the cytoplasmic side of the membrane. It is lined exclusively by hydrophobic residues, with exceptions of the conserved H157 and H305. Figure taken from reference [3]. **Right**: Ammonium recruitment site and substrate channel of _A. fulgidus_ Amt-1 protein. Model of a NH_4^+ molecule in the recruitment site (top view from the extracellular side) presumably being selected by its hydrogen bonding to the side chain of S208 and cation-π interaction with the side chain of W137.

To test the hydrophobic nature of the pore, a hydrophobic gas, xenon, was pressurized into the crystals of Amt-1 (Figure 3.7). Two out of 15 xenon atoms per monomer were found in the lumen of the pore, emphasizing the hydrophobicity of the substrate channel with the exception of two histidine residues [3]. The side chains of these residues (H157 and H305), both highly conserved among the entire Amt family, lie adjacent, with their imidazole ring edges facing the cavity such that an hydrogen bond is formed between their δ nitrogens [3] (Figure 3.6). Their high conservation, and localization deeply buried in the hydrophobic pore lead to the proposal of a functional role for this pair in substrate deprotonation/reprotonation. This hypothesis is, however, still lacking experimental evidences to support it.

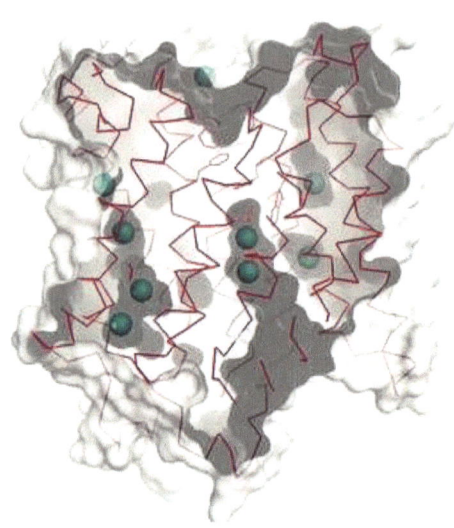

Figure 3.7: Xenon binding sites in *A. fulgidus* Amt-1 protein
The surface representation shows the substrate channel and several internal cavities as well as the position of 9 out of a total of 15 Xe atoms located in the monomer structure. Two Xe sites have been found to occupy a large hydrophobic pocket in the structure. Figure taken from reference [3].

3.4. Mechanistic consideration

Despite the large number of Amt proteins being identified and some of them even isolated, the mechanism of ammonium movement across the plasma membrane mediated by these proteins is still unclear. Several models of ammonium uptake are discussed, but none of them can explain exactly the transport pathway in detail.

One model suggested a secondary active transport system with a co-transport of NH_3/H^+ [41]. This transport system would be electrogenic and dependent on the proton motive force as well as on the membrane potential of the cell. This model was later confirmed in whole cells uptake measurements using ^{14}C-labelled methylammonium on Mep proteins from yeast [13, 14] and Amt proteins from *Arabidopsis thaliana* [42, 12]. Voltage clamp experiments performed with the tomato protein *Le*Amt1;1 [43] and the RhBG glycoprotein [44] reconstituted in *Xenopus laevis* oocytes, made possible to show active NH_4^+ uptake, dependent on the membrane potential. These measurements showed an increase in the ammonium uptake as the external ammonium concentration was increased. This model, however, was not able to distinguish between symport of NH_3 with H^+ or uniport of NH_4^+, because both processes show an identical net charge in these experiments.

A further model suggested Amts as ammonia gas channels [45, 46]. *In vivo* experiments with whole cells showed that the Amt proteins could act as mediators for the permeation of NH_3, when the external concentration of NH_3/NH_4^+ was low ($\leq$ 50nM). This hypothesis was supported by *in vitro* assays with purified AmtB reconstituted into proteoliposomes [1]. The pH variation inside proteoliposomes was monitored via the fluorescence change of a pH-sensitive dye (carboxyfluoresceine). Experiment showed a significant internal alkalinisation rate for AmtB containing proteoliposomes (interpreted as due to an NH_3 influx), which confirmed structure-based conclusion that AmtB is an NH_3 channel rather than an NH_4^+ transporter. It is important to mention that in this experiment an AmtB variant carrying three undescribed substitutions (F68S, S126P, and K255L) was used eventually providing a strong reason for the different results obtained by another group using wild type AmtB [47].

But this theory has a very fundamental, bioenergetics problem. Namely, at physiological pH values ammonium will be almost in the cationic form, such that translocation of NH_3 requires extracellular deprotonation and intracellular reprotonation. This would, in sum, lead to an antiport of NH_4^+ with H^+ and as a consequence, create a problem for an organism with a proton gradient for energy conservation since this H^+ counter flow would result in a reduction of the membrane proton motive force and thus necessarily require energy in the form of adenosine threephosphate (ATP).

A possible explanation if deprotonation occurs is the following: deprotonation of the bound NH_4^+ at some, still unknown, site would result in the conductance of NH_3 and H^+ in two separate but mechanistically coupled pathways leading to a net NH_4^+ uniport [16]. This model is also consistent with the presence of a proton motive force over physiological cell membranes.

3.5. Regulatory aspects of ammonium transport activity/sensing

Upon import into the cytoplasm, ammonium has to be incorporated into glutamate or rapidly redistributed to other subcellular compartments since high concentrations of NH_4^+ in the cytosol are toxic [48]. The glutamine sythetase/glutamate synthase (GS/GOGAT) system is the main assimilatory pathway for ammonium in bacteria and plants. This pathway is highly regulated at both transcriptional and posttranslational levels. In enteric bacteria, GS represents, biochemically and genetically, a well-studied nitrogen metabolism enzyme.

$$NH_3^+ + {}^+H_3N-\underset{\underset{\underset{\underset{C}{|}}{CH_2}}{\overset{\overset{COO^-}{|}}{\underset{|}{CH}}}{} + ATP \longrightarrow {}^+H_3N-\underset{\underset{\underset{C}{|}}{CH_2}}{\overset{COO^-}{\underset{|}{CH}}} + ATP + Pi$$

Equation 3.1: Glutamine synthetase (GS) requires energy in the form of ATP for ammonia incorporation, using glutamate as an acceptor. Picture taken from reference [49].

$$\text{(2-oxoglutarate)} + \text{(glutamine)} \;\rightleftharpoons\; \text{(glutamate)} + \text{(glutamate)}$$

Equation 3.2: Glutamine can be a precursor for the synthesis of glutamate, with the reaction of glutamate synthase, also known as GOGAT (glutamine: 2-oxyglutarate aminotransferase). Picture taken from reference [49].

In this context, GlnK and GlnB (P_{II} proteins) are key elements in the regulation of nitrogen assimilation in prokaryotes and plants. They integrate metabolic status information by monitoring and responding to the cellular levels of 2-oxoglutarate (OG) and glutamine as the substrate and product of ammonium assimilation through glutamine synthetase (GS) and the NADPH-dependent glutamine: 2-oxoglutarate aminotransferase (GOGAT) system [50] as well as to the ratio of ATP and adenosine diphosphate (ADP) as an indicator of the cellular energy level. The GlnK protein is known to have some well defined intracellular targets in eubacteria, and GlnK binding to AmtB may also constitute a component of the regulatory process in these systems. They include the nitrogen fixation regulatory proteins (NifLA), which are controlled by GlnK in *Klebsiella pneumoniae* and *Azotobacter vinelandii* [51, 52] and ADP-ribosylation of nitrogenase in *Rhodospirillum rubrum*, which is regulated by the GlnK homologue GlnJ [53].

The data obtained suggested that GlnK binds to AmtB when the cellular nitrogen status reaches a certain level, so as to inhibit further ammonium transport [36]. This model required that GlnK binding to AmtB should be sensitive, rapid and reversible, involving the transporter as an integral part of the signal transduction cascade and thus acting as an ammonium sensor.

Observations from crystallography studies with *E. coli* [54, 55] and electron microscopy studies with *Methanococcus jannaschii* complexes [56] confirmed the previously established structural model initially proposed for the *A. fulgidus* GlnK-Amt complex [3]. This complex is formed such that the T-loops of GlnK extend into the substrate exit channels of the trimeric Amt protein and prevent further uptake of NH_4^+/NH_3 [54, 55, 56]. The levels of nitrogen in the cell trigger this interaction such that complex formations is blocked when intracellular nitrogen is abundant and the uptake of potentially cytotoxic ammonium for the cell.

4. Materials and Methods

4.1. Materials

All chemicals used were of analytical purity grade and have been obtained from the companies: Applichem (Darmstadt, Germany), Merck (Darmstadt, Germany), Roth (Karlsruhe, Germany), Sigma-Aldrich (Deisenhofen, Germany) and Anatrace (Maumee, USA).

Protein molecular weight determination in SDS-PAGE was achieving by comparison with molecular weight markers from Fermentas (Unstained Protein Molecular Weight Marker and Page Ruler Plus Prestained).

Restriction endonucleases, T4 DNA ligase, were all from Fermentas. *Pfu* DNA Polymerase used for PCR was also from Fermentas. *Pfu* Turbo DNA Polymerase used for site-directed mutagenesis was from Strategene. pET-21a expression vector was from Novagen.

4.1.1. Bacterial strains

XL10-Gold   *[Tet^{R}Δ(mcrA)183 Δ(mcrCB-hsdSMR-mrr)173 endA1 supE44 thi-1 recA1 gyrA96 relA1* lac *Hte* [F' *proAB lacIqZΔDM15 Tn10* (TetR) *Amy CamR*], source Strategene.

BL21(DE3)   *[E. coli B F⁻ dcm ompT hsdS (r$_B^-$ m$_B^-$) gal λ (DE3)]*, source Strategene.

C43 (DE3)   is derived from the BL21 (DE3) strains. This strain has at least one uncharacterized mutation which enables effective overexpression of membrane proteins preventig cell death associated with the expression of toxic recombinant proteins. The strain C43(DE3) was further derived from C41(DE3) transformed with the F-ATPase subunit gene and it contains no plasmid [57].

4.1.2. Growth Media

4.1.2.1. Lysogeny broth (LB) medium

LB medium is a nutritionally rich medim primarily used for the growth of bacteria. This medium is widely used in molecular microbiology applications for the preparation of plasmidic DNA and recombinant proteins. It is one of the most common using media for maintaining and cultivating recombinant strains of *E. coli.*

For LB medium preparation, 10 g of Tryptone (final conc. 1 %), 5 g of yeast extract (final conc. 0.5 %) and 10 g NaCl (final conc. 1 %) were dissolved in destiled water (ddH$_2$O) to perform a total volume of 1 litre. This medium was then sterilised by autoclaving at 121 °C for 20 min and stored at 4 °C.

For preparing LB agar plates, sterile LB agar medium (1 % Tryptone, 0.5 % yeast extract, 1 % NaCl and 1.2 % Agar-Agar) is heated to liquify the agar (around 50 °C). At this stage antibiotics are added, if required and the liquid is dispensed into sterile Petri dishes (~5 mL per dish) and let cool down at room temperature until it becomes solid again. This process takes up to 30 min, upon which the petri dishes are turned upside down to avoid condensation on the agar. Plates are then stored closed at 4 °C.

4.1.2.2. M9 Minimal medium

Minimal media contain the minimum nutrients necessary for bacteria to growth. It typically contains a carbon source such as glucose or a less energy-rich source like succinate, various salts and essential elements such as magnesium, nitrogen, phosphorus and sulfur that can be adjusted for different bacterial species and growth conditions, plus any other substance that, in resume, is required for the bacteria to synthesize proteins and nucleic acids for its normal growth.

Table 4.1. Composition of M9 Minimal medium

Conc. (mM)	Component
42.25	Na$_2$HPO$_4$ x 7H$_2$O
22.06	KH$_2$PO$_4$
8.5	NaCl
1	MgSO$_4$
22	glucose
0.1	CaCl$_2$

4.2. Methods

4.2.1. Creation of a ΔamtB C43 (DE3) *Escherichia. coli* strain

The TargeTron™ Gene Knockout System (Sigma-Aldrich) is based on rapid and specific disruption of bacterial genes by insertion of group II introns. Group II introns insert themselves via the activity of an RNA-protein complex (RNP) expressed from a single plasmid provided in the kit. The RNA portion of the RNP is easily mutated to re-target insertion into a user-specified chromosomal gene.

Figure 4.1: TargeTron™ Gene Knockout System. Picture taken from reference [58].

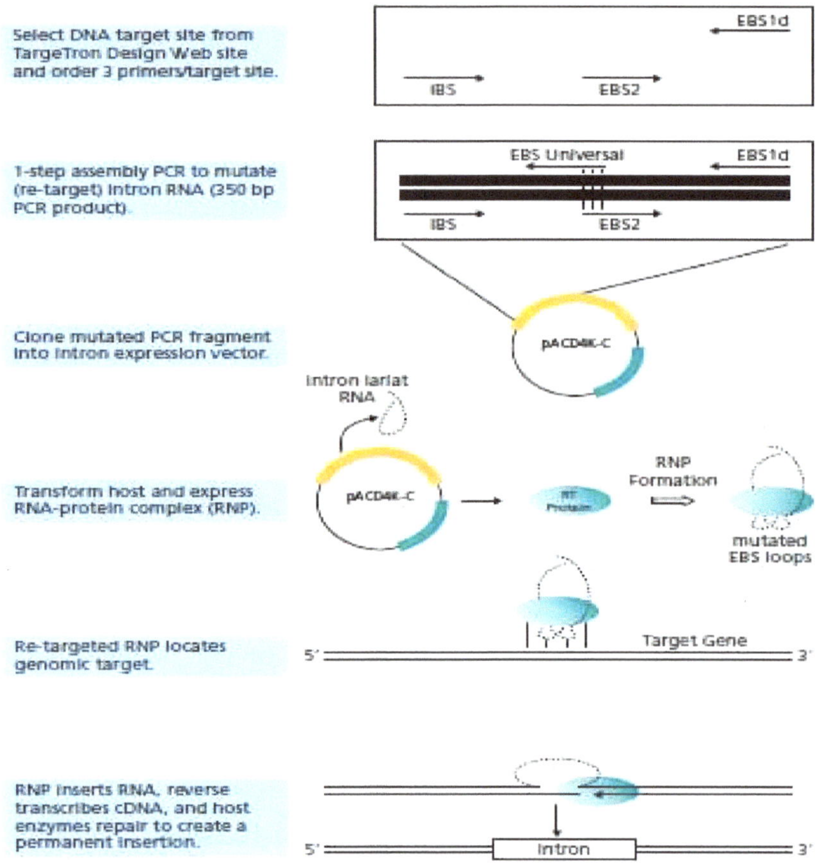

Figure 4.1 shows the process of using TargeTron system to knockout bacterial genes. First, a computer algorithm is used to identify target sites in the gene of interest (in our work it was *amtB* from *E. coli*). A typical 1 kb gene can be expected to contain 5 to 11 group II intron insertion sites. Second, the computer algorithm outputs primer sequences, which are used to mutate (re-target) the intron by PCR. Next, the mutated 350 bp PCR fragment is ligated into a linearized pACD4K-C vector that contains the remaining intron components. The ligation

reaction is transformed into the host followed by expression of the re-targeted intron. *ΔamtB* C43 (DE3) *E. coli* strains are then selected using an ampicillin marker that is activated upon chromosomal insertion. Using gene specific primers, ampicillin resistent colonies are PCR screened [59].

Table 4.2: Primers used for making the *ΔamtB* C43 (DE3) *E. coli* strain

IBS	5-AAAAAAGCTTATAATTATCCTTATGCGACGTCTTCGTGCGCCCAGATAGGGTG-3'
EBS1d	5'-CAGATTGTACAAATGTGTGATAACAGATAAGTCGTCTTCGGTAACTTACCTTTCTTTGT-3'
EBS2	5'-TGAACGCAAGTTTCTAATTTCGGTTTCGCATCGATAGAGGAAAGTGTCT-3'

4.2.2. Colony PCR

This procedure uses PCR to screen for successful ligations from a single colony on transformation plate without plasmid preparation. A colony is taken from LB agar plate and resuspended in PCR tube containing appropriate amount of ddH$_2$0 following the PCR reaction mixture (Table 4.3).

Table 4.3: PCR reaction mixture

1X	*Pfu* buffer
0.2 mM	dNTPs
2.5 ng/μL	Forward primer
2.5 ng/μL	Reverse primer
0.04 U/μL	*Pfu* DNA Polymerse

Table 4.4: PCR thermocycler program

98 °C	3 minutes
98 °C	30 seconds
60 °C	1 minute
72 °C	3 minutes
72 °C	10 minutes
~4 °C	pause

4.2.3. Examination of concentration and purity in DNA preparations

The most comprehensive way to evaluate DNA concentration and purity is to use both ultraviolet-visible spectrophotometeric measurements and agarose gel eletrophoresis.

DNA strongly absorbes light at 260 nm and therefore its concentration can be easily quantified by the sample absorbance at this wavelength, according to the Lambert-Beer law [60]. Since tyrosine and tryptophan residues have an absorption maxima at 280 nm, this absorbance (A_{280}) is commonly used as an indicator of protein contamination in DNA samples. Consequently, a good quality DNA sample has an A_{260}/A_{280} ratio between 1.8-2.0. A lower absorption ratio indicates that the preparation is contaminated with proteins and aromatic substances (e.g., phenol), while a higher ratio indicates a possible contamination with RNA [61].

A disadvantage of using solely spectrophotometric measurements is that contaminants such as genomic DNA, RNA, and proteins - that still may display some absorbance at 260 nm - will also contribute to the intensity at this absorbance and thus lead to an overestimation of the DNA concentration. Agarose gel electrophoresis is therefore used in combination allowing to observe the overall quality of the DNA sample preparation. In this method, a quantitative dye such as ethidium bromide, allows for DNA visualization while it can still be used as an alternative approach to evaluate the DNA concentration since it is not affected by the above mentioned contaminants. Contaminating RNA or genomic DNA can be detected on an agarose gel, since RNA will run as a low molecular weight smear and genomic DNA as a high molecular band.

4.2.4. Agarose Gels

The simplest and most effective method for separating and identifying fragments of DNA is by running an agarose gel electrophoresis. Most agarose gels are made between 0.7 % and 2 %. A 0.7 % gel will show good separation (resolution) of large DNA fragments (5–10 kb) and a 2 % gel will show good resolution for small fragments (0.2–1 kb). Polymerisation of agarose in a defined buffer system provides a matrix of defined average pore size depending on the concentration of agarose. This allows for separation of fragments reflecting their mutual sizes, as smaller molecules migrate faster through the gel matrix than lager molecules do. DNA markers allow for size estimation of observed bands. Gels were made using TAE buffer (Table 4.5) with 0.8 % agarose. Samples were mixed with 1X loading dye (Table 4.6). Electrophoresis procedure was carried out in a horizontal gel system and run at 80 V.

Table 4.5 : Composition of the 50X TAE Buffer

0.4 M	Tris Base
0.4 M	Glacial Acetic Acid
10 mM	EDTA

Table 4.6: Composition of the 6X loading dye

10 mM	Tris-HCl pH 7.6
0.0 3 %	bromphenol blue
0.0 3 %	xylene cyanol FF
60 %	glycerol
60 mM	EDTA

4.2.5. Site-directed mutagenesis

Site-directed mutagenesis is a simple, rapid and efficient method to perform point mutations, replace, delete or insert one or several amino acid residues using a thermal cycling technique in combination with Dpn I digestion of parental DNA.

The basic site-directed mutagenesis reaction entails mixing a circular plasmid, two complementary primers, and a DNA polymerase together in the presence of an appropriate buffer. Site-directed mutagenesis method is performed by using *Pfu* Turbo DNA Polymerase. This polymerase replicates both plasmid strands with high fidelity and without displacing the mutant oligonucleotide primers. Heat is applied to denature the plasmid, temporarily separating both strands of the DNA. The heat is gradually lessened allowing the primers to pair with a similar sequence in the single stranded plasmid. At some point, the polymerase will recognize an appropriately primed oligo-plasmid pair and create a new DNA fragment extending off the primed oligo. This new fragment is identical to the circular plasmid and contains the added oligo and any mutations engineered into it. A variety of steps are applied to remove and or modify the original plasmid. The Dpn I endonucleases (target sequence: 5'-Gm6ATC-3') is specified for methylated and hemimethylated DNA and is used to digest the parental DNA template and to select for mutation-containing synthesized DNA. As most *E. coli* strains produce methylated DNA, they are not resistant to Dpn I digestion.

The reaction mixture is then introduced into bacteria, which, at some point, will convert the newly created linear DNA fragments into circular plasmids. If an individual bacterium is isolated and allowed to grow under selection for plasmids, each will contain a unique plasmid inherited from the site-directed mutagenesis reaction. Some, but not all, will contain the modified plasmid.

This protocol requires a small amount of the original DNA (about 300 ng) plus a forward and reverse primers of about 36 bp, containing the desired mutation in the middle (about 125 ng). dNTPs are added to a final concentration of 0.2 mM. Lastly, 0.32 µL of *Pfu* Turbo Polymerase (2.5 U/µL) is added and the reaction mixture is cycled in a PCR thermal cycler. The number of cycles depends on the procedure (e.g. 12 cycles for a point mutation; 16 for a single amino acid change, 18 for a multiple amino acid deletions or insertions) (see Table 4.7) [62].

Temperature	Time	Nr. of cycles
95 °C	30 sec	1
95 °C	30 sec	
55 °C	60 sec	
68 °C	7 min	16
10 °C	pause	

Table 4.7: PCR program used for the mutagenesis of *Af* Amt-1

After this step, 1 µL of Dpn I (10 U/µL) restriction enzyme is added to each reaction and incubated at 37 °C for 1 h. After digestion, 1 µL of this solution is transformed into a 50 µL aliquot of XL-10 Gold or XL-1 Blue chemically competent cells (see section 4.2.6) and plated on LB agar Petri dishes. The following day one colony is picked and inoculated in 5 mL of LB medium supplemented with 100 µg/mL ampicilline and let grow at 37 °C for 10-12 hours. Isolation of plasmid DNA was performed with the Zyppy™ Plasmid Miniprep Kit (Zymo Research).

The desired mutation was then confirmed by sequencing (GATC BIOTECH) using 30 ng/µL of purified plasmid DNA. Results were analyzed using Chromas (Technelysium).

Table 4.8: List of mutations and respective primers used to create them in *Af*Amt-1. Designed mutations are highlighted in red.

Mutant	Primer sequence (F-forward, R-reverse)
H157F	(F) 5'- GCTGGAGGTATGGTTGTTTTCATAAGCTCGGG-3'
	(R) 5'-CCCGAGCTTATGAAAACAACCATACCTCCAGC-3'
H305E	(F) 5'-GCTTGGGCGATGAACGGAATAGGCGGTTTATGGGG-3'
	(R) 5'-CCCCATAAACCGCCTATTCCGAAAATCGCCCAAGC-3'
D149N	(F) 5'-CAAAGCTCGGCGCCCTCAATTTTGCTGGAGGTATG-3'
	(R) 5'-CATACCTCCAGCAAAATTGAGGGCGCCGAGCTTTG-3'
D149E	(F) 5'- CAAAGCTCGGCGCCCTCGAGTTTGCTGGAGGTATG-3'
	(R) 5'- CATACCTCCAGCAAACTCGAGGGCGCCGAGCTTTG-3'
D149L	(F) 5'-CAAAGCTCGGCGCCCTCCTGTTTGCTGGAGGTATG-3'
	(R) 5'-CATACCTCCAGCAAACAGGAGGGCGCCGAGCTTTG-3'
H157A	(F) 5'-GTATGGTTGTTGCAATAAGCTCG-3'
	(R) 5'-CGAGCTTATTGCAACAACCATAC-3'
H157D	(F) 5'-GTATGGTTGTTGATATAAGCTCG-3'
	(R) 5'-CGAGCTTATATCAACAACCATAC-3'
A258T	(F) 5'-GCTGGGCTTACCGCCATAAC-3'
	(R) 5'-GTTATGGCGGTAAGCCCAGC-3'
A258V	(F) 5'-GCTGGGCTTGTGGCCATAAC-3'
	(R) 5'-GTTATGGCCACAAGCCCAGC-3'
W137H	(F)5'-CCTTCGCACACTGGCTTCACGGTGGGGGGTGGCTGGC-3'
	(R)5'-GCCAGCCCCCCCACCGTGAAGCCAGTGTGCGAAGG-3'
V156L	(F) 5'-GGTATGGTTCTTCACATAAGC-3'
	(R) 5'-GCTTATGTGCAGAACCATACC-3'

V156S	(F) 5'-GGTATGGTTAGCCACATAAGC-3'
	(R) 5'-GCTTATGTGGCTAACCATACC-3'
V156A	(F) 5'-GGTATGGTTGCCCACATAAGC-3'
	(R) 5'-GCTTATGTGGGCAACCATACC-3'
W201A	(F) 5'-CCTGCTTTGGTTTGGGGCGTTCGGATTCAACGG-3'
	(R) 5'-CCGTTGAATCCGAACGCCCCAAACCAAAGCAGG-3
W201F	(F) 5'-CCTGCTTTGGTTTGGGTTTTTCGGATTCAACGG-3'
	(R) 5'-CCGTTGAATCCGAAAAACCCAAACCAAAGCAGG-3'
A112L	(F) 5-ATTGCTGAGAGACTTAAAGTTTCATCG-3'
	(R) 5'-CGATGAAACTTTAGGTCTCTCAGCAAT-3'
H157E	(F) 5'-GGAGGTATGGTTGTTGAAATAAGCTCGGGATTTGC-3'
	(R) 5'-CGAAATCCCGAGCTTATTTCAACAACCATACCTCC-3'
H305E	(F) 5'-CTTGATGCTTGGGCGATTGAAGGAATGGCGGTTTATG-3'
	(R) 5'-CATAACCCCTATTCCTTCAATCGCCCAAGCATCAAG-3'
I304E	(F) 5'-GCCTTGATGCTTGGGCGGAACACGGAATAGGCGG-3'
	(R) 5'-CCGCCTATTCCGTGTTCCGCCCAAGCATCAAGGC-3'
N205L	(F) 5'-GGGTGGTTCGGATTCCTTGGCGGAAGTGC-3'
	(R) 5'-GCACTTCCGCCAAGGAATCCGAACCACCC-3'
T261V	(F) 5'-GGGCTTGCCGCCATAGTCCCCGCAGCAGGC-3'
	(R) 5'-GCCTGCTGCGGGGACTATGGCGGCAAGCCC-3'
W137L	(F) 5'-GCCCCCTTCGCACACTTGCTT TGGGGTGGGGGG-3'
	(R) 5'-CCCCCCACCCCAAAGCAAGTGTGCGAAGGGGGC-3'
D149A	(F) 5'-GCAAAGCTCGGGCCCTCGCCTTTGCTGGAGGTATG-3'
	(R) 5'-CATACCTCCAGCAAAGGCGAGGGCCCGAGCTTTGC-3'

4.2.6. Transformation of *Escherichia coli* competent cells

The purpose of this technique is to introduce a foreign plasmid into bacteria and to use those bacteria to amplify the plasmid in order to make large quantities of it. This is based on the natural function of a plasmid: to transfer genetic information vital to the survival of the bacteria. A plasmid is a small circular DNA (2000-10000 bp) that contains important genetic information for the growth of bacteria. In nature, this information is often for a gene encoding a protein that will make the bacteria resistant to an antibiotic (usually ampicillin). In practical terms, a gene of interest is inserted into a plasmid and this newly constructed plasmid is then inserted into *E. coli* cells that can be, or not, sensitive to a particular antibiotic. The bacteria are then spread over a plate that contains that antibiotic providing selective pressure because only bacteria that have now acquired the plasmid can grow on the plates.

In practice, chemically competent cells of *E. coli* are transformed with plasmidic DNA by adding 75-100 ng of it to cells freshly thawed on ice. Cells and plasmid are incubated for 30 min on ice and heat shocked at 42 °C for 30 seconds, before chilling the cells again for 2 min on ice. By adding 200 μL of LB and incubation this solution at 37 °C, cells are becoming able to express ampicillin resistance markers encoded by the newly incorporated plasmidic DNA.

4.2.7. Heterologous overexpression in *Escherichia coli* C43 (DE3) competent cells

Heterologous overexpression refers to the process of introducing the gene of a particular protein of interest in a cell host which does not normally make that protein. Initially, the protein gene is cloned into an expression vector such as pET-21a and then host cells (chemically competent) are transformed with that plasmid (see section 4.2.6).

In order to optimize the expression level of Amt-1 variants, expression tests varying concentration of IPTG and temperatures were made.

In practice, one fresh *E. coli* C43(DE3) colony of chemically transformed cells with the desired plasmid carrying mutation of interest was picket out of an LB agar plate and inoculated into 500 mL of LB-medium, supplemented with 100 μg/mL ampicilline and let grow overnight at 37 °C with 450 rpm stirring. The following day, 20 mL of the overnight pre-culture were taken to inoculate a fresh 500 mL of LB-medium, supplemented with 100 μg/mL ampicilline. Cell growth was monitored by measuring the optical density at 600 nm (OD_{600nm}) in culture samples of 1 mL, taken at chosen growth time points. Induction was initiated with 0.5 or 1 mM isopropyl β-D-1-thiogalactopyranoside (IPTG) at OD_{600nm} 0.6-0-8. Once good expression conditions were obtained, up scaling of protein expression was done in 10-15 L cultures.

Cell growth was stopped by harvesting the cell cultures at 4 000 g, for 10 min at 4 °C (Rotor JLA-8.100). The cell pellet was collected using a spatula, the total wet cell mass was determined and immediately shock frozen in liquid nitrogen for storage at -20 °C until further use for protein purification.

4.2.8. Cell disruption

Frozen cells were resuspended in cold buffer A (20 mM Tris-HCl pH 8.0, 300 mM NaCl and 10 % glycerol) at a ratio of 3 mL buffer A per gram of wet cell weight. At this stage, buffer A was supplemented with a protease inhibitor cocktail mixture (Roche) to inhibit cytosolic serine, cysteine and other metalloproteases. One protease pill was added per 50 mL of buffer. Disruption of cells was carried out using a French Press (SLM Amicon French® Pressure Cell) or a Fluidizer (Microfluidics, M-110P). In both methods, the cell suspension at the central bore of a steel cylinder cell is put under a mechanical pressure of about 1100 psi. There is only one way out, via a small Teflon ball attached to a screw valve. Upon release, bacterial cells are exposed to such pressure drop that they instantaneously burst.

4.2.9. Preparation of Membranes

After cell disruption, cell debris is discarded by a centrifugation step of 15 min at 30 000 g at 4 °C (rotor JA-30.50). A further ultracentrifugation step, at 350 000 g (rotor 70 Ti) carried out for 60 min at 4 °C, allowed for separating the membranes from the cytoplasm. The membrane fraction was then resuspended in buffer A (10 mL per gram of membranes).

4.2.10. Solubilisation of Membrane Proteins

In order to understand the structure and function of membrane proteins, it is necessary to carefully isolate these proteins in their native form and in a highly purified state. For this purpose we use detergents. Detergents are amphipathic molecules that contain a polar group (head) at the end of a hydrophobic carbon chain (tail). Due to their amphipathic character, they possess unique properties in water. While their polar group forms hydrogen bonds with water molecules, the hydrocarbon chains aggregate due to hydrophobic interactions. These properties allow detergents to be soluble in water by forming spherical structure called micelles.

Solubilisation is a process that can be divided into different stages. At low concentration, detergents bind to the membrane by partitioning into the lipid bilayer. At higher concentrations, when the bilayers are saturated with detergents, the membranes disintegrate to form mixes micelles with detergents molecules. In the detergent-protein micelles, hydrophobic regions of the membrane proteins are surrounded by the hydrophobic chains of micelles. In the final stages, the solubilisation of membranes leads to the formation of mixed micelles consisting of lipids and detergents and detergents micelles containing proteins.

The critical micelle concentration (CMC) is the detergent concentration above which micelles form. Bellow CMC, the detergent molecules in solution exist as monomers. For solubilisation, the detergent concentration has to be higher than the CMC value, because the membrane proteins must be able to incorporate into micelles.

For solubilization, n-dodecyl-β-D-maltopyranoside ($D_{12}DM$; CMC=0.0087 %) was added drop wise, to reach a final concentration of 1 % (w/v) while stirring the membrane solution at

4 °C. Centrifugation at 350 000 g for 45 min at 4 °C was then used to separate the solubilized from the non-solubilized fractions. The solubilized fraction was then used for Ni-affinity chromatography.

4.2.11. Ni-Affinity Chromatography

Ni-Affinity Chromatography uses the ability of histidine to bind metals. Nickel for example is bound to an agarose bead by chelation using nitroloacetic acid (NTA) beads. Bound proteins (that is, proteins carrying a N- or C-terminus His tag) can be selectively eluted with imidazole because this molecule competes with the bound protein for the occupied places in the Ni-atoms.

The hexa-histidine affinity tag of the recombinant Amt-1 proteins overexpressed in pET-21a can chelate to sepharose-matrix coordinated Ni^{2+} ions. Consequently, purification of these proteins was done by affinity chromatography at 20 °C, using an Äkta prime chromatographic system (GE Healthcare) where protein elution was monitored at 280 nm. The $D_{12}DM$ solubilized membrane fraction was loaded onto a nickel HiTrap FF 1 mL column (GE Healthcare), pre-equilibrated with 5 column volumes of buffer A supplemented with 0.03 % $D_{12}DM$. After sample loading and washing (back to baseline) in buffer A, a detergent exchange step in buffer A from 0.03 % $D_{12}DM$ to 0.05 % n-dodecyl-N, N-dimethylamine-N-oxide (LDAO; CMC=0.023 %) was done. Consequent removal of unspecifically bound proteins from the HiTrap column material was performed by a stepwise increase of the imidazole concentration in buffer B (20 mM Tris-HCl pH 8.0, 300 mM NaCl and 10 % glycerol, 0.5 M imidazole pH 8.0) to 85 mM. After washing all eluted proteins with 85 mM imidazole-containing buffer, another step increase in buffer B to 250 mM imidazole eluted the Amt-1 protein. Fractions containing Amt-1 were pooled and concentrated to a final volume of about 500 μL in an ultrafiltration Vivaspin concentrator (Sartorius) ran at 3600 g in an Eppendorf 5810 R centrifuge.

4.2.12. Size Exclusion Chromatography

Size exclusion chromatography separates molecules on the basis of their molecular size, through a highly porous gel matrix (dextran, agarose or polyacrylamide). Separation effect depends on the porous size of the matrix material and the molecular weight of the proteins.

Bigger proteins often elute early, whereas smaller proteins are able to move into the interstitial spaces and porous of the matrix, and thus have a much lower speed throughout the column. It is the reason why smaller proteins elute later from such columns.

A HiTrap^{TM} desalting column with Sephadex™ G-25 Superfine resin was consequently used to exchange the buffer of the protein sample to buffer C (20 mM Tris-HCl pH 8.0, 100 mM NaCl, 10 % glycerol, 0.05 % LDAO). Protein samples were collected and concentrated to a final volume of about 100-150 µL in an ultrafiltration Vivaspin concentrator (Sartorius) ran at 3600 g in an Eppendorf 5810 R centrifuge. Protein concentration was determined (see section 4.2.15) and aliquoted in 25-50 µL before shock frozen in liquid nitrogen and stored at -20 °C.

4.2.13. Sodium Dodecyl Sulphate-Polyacrylamide Gel Electrophoresis

Polyacrylamide gel electrophoresis (PAGE) is performed for size analysis of protein samples and for evaluation of protein purity, as well. Sodium dodecyl sulfate (SDS) is an anionic detergent that binds to proteins (1.4 g SDS per gram of protein in solutions of 1 % [63]) and in this way introduces a global negative charge through its polar head groups. The gel matrix is made of polyacrylamide. The polyacrylamide chains are crosslinked by N, N-methylene bisacrylamide comonomers. Polymerisation is initiated by ammonium persulfate (radical source) and catalysed by TEMED (a free radical donor and acceptor). A matrix of defined average pore size, causes size separation of proteins during migration through an electric field. Besides SDS used in standard SDS-PAGE buffers, mercaptoethanol or other reducing reagents (e.g. dithiothreitol) are generally useful to break the protein tertiary structure stabilizing disulfide bridges. Gels were prepared in the Minive Complete vertical electrophoresis system (Amersham Biosciences, Germany) using 12.5 % and 4 % of acrylamide/bisacrylamide for the separating and stacking gels, respectively (Table 4.9).

Table 4.9: Composition of separating and stacking gel

Separating gel	Stacking gel
30 % Bis-acrylamide	30 % Bis-acrylamide
1.88 mM Tris-HCl, pH 8.0	0. 625 mM Tris-HCl, pH 6.8
0.5 % SDS	0.5 % SDS
TEMED	TEMED
10 % APS	10 % APS

Protein samples were mixed with 1 μL of loading buffer (Table 4.10) before being applied to the gel. A mixture of protein molecular weight markers was also applied as a size standard. The running chamber was immersed in running buffer (Table 4.10) and attached to a power supply unit to run at 25 mA per gel with 300 V.

Table 4.10: Composition of loading and running buffers for SDS-PAGE

5X loading buffer	1X running buffer
0.5 M Tris-HCl pH 6.8	25 mM Tris base
40 % glycerol	192 mM glycine
8 % SDS	0.1 % SDS
0.004 % bromphenol blue	

4.2.14. Coomassie staining

For the staining of SDS-PAGE gels, a coomassie-based staining solution was made with two coomassie dyes: the G-250 ("G" for "greenish") and R-250 ("R" standing for "reddish"). These dyes bind unspecifically to cationic and nonpolar, hydrophobic amino acid residues of the protein and allow to visualise them as a blue band. The acid/ethanol mixture fixes the proteins in the gel and prevents it from moving or being washed out during the staining [63, 64]. Gels were soaked in 50 mL of Commasie staining solution (Table 4.11) for about 1 h at room temperature and then eluted with a destaining solution over night (Table 4.11).

Table 4.11: Composition of Commasie staining and destaining solution

Staining solution	Destaining solution
10 % ethanol	
5 % acetic acid	10 % ethanol
0.002 % Commasie (G250/R250=4:1)	

4.12.15. Bicinchoninic acid protein assay (BCA)

The BCA Protein Assay combines the well-known reduction of Cu^{2+} to Cu^{1+} by proteins in an alkaline medium with the highly sensitive and selective colorimetric detection of the cuprous cation (Cu^{1+}) by bicinchoninic acid. As stated in reference 65: ''The first step is the chelation of copper with protein in an alkaline environment to form a light blue complex. In this reaction, known as the biuret reaction, peptides containing three or more amino acid residues form a colored chelate complex with cupric ions in an alkaline environment containing sodium potassium tartrate. In the second step of the color development reaction, bicinchoninic acid (BCA) reacts with the reduced (cuprous) cation that was formed in step one. The intense purple-colored reaction product results from the chelation of two molecules of BCA with one cuprous ion. The BCA/copper complex is water-soluble and exhibits a strong linear absorbance at 562 nm with increasing protein concentrations''. The BCA reagent is approximately 100 times more sensitive (lower limit of detection) than the pale blue color of the first reaction.

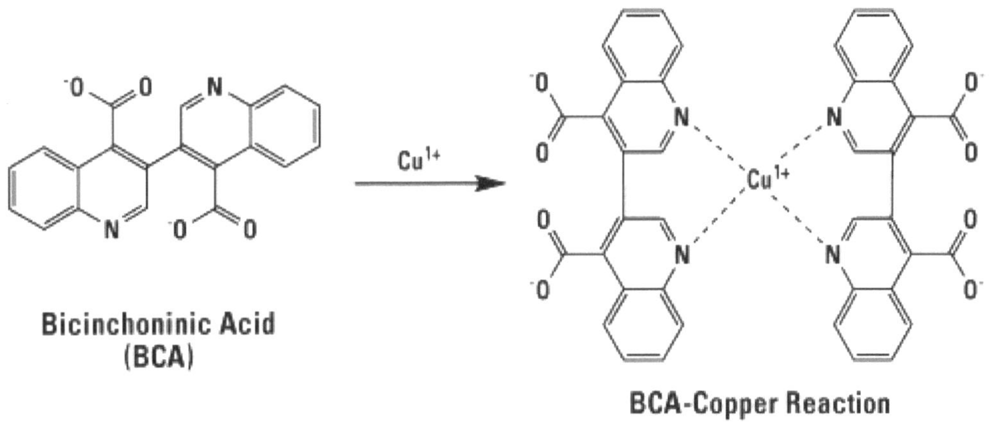

Bicinchoninic Acid (BCA)

BCA-Copper Reaction

Figure 4.2: **The reaction of BCA with cupric ion.** Two molecules of BCA bind to each molecule of copper that had been reduced by a peptide-mediated biuret reaction. Picture taken from reference [66].

A standard protein, such as bovine serum albumin (BSA) is used in a convenient range for the establishment of a calibration curve (0 – 25 mg/mL). This test is quick and sensitive. The sensitivity of this method is high and protein amounts down to 0.5 μg/mL protein are easily detected [65]. The reaction takes place in an alkaline environment in which almost all proteins remain in solution. A disadvantage of this method is coming from the effect of some chemicals (EDTA, ammonium sulphate, N-acetyl-glucosamine, glycine, reducing materials such as glucose, DTT or Sorbitol; and a host pharmaceuticals such as chlorpromazine, penicillin and vitamin C [66]) on particulary compounds from BCA kits.

4.2.16. Western blot

Western blotting is a technique used to specifically identify and locate proteins on a membrane based on their capacity to bind specific antibodies. It uses gel electrophoresis to separate denatured proteins by the length of their polypeptide chain (denaturing conditions, e.g. SDS-PAGE) or by the tertiary/quaternary structure of the protein (native, non-denaturing conditions, e.g. in native gels). The proteins separated this way are then transfered onto a membrane (typically nitrocellulose or polyvinylidendifluoride (PVDF) by an electric current (electroblotting). Following electrophoresis, a wet or semi-dry blotting transfer system is set-up; a stack is assembled in the following order from cathode to anode: sponge | filter paper soaked in transfer buffer (Table 4.12) | gel | PVDF membrane | filter paper soaked in transfer buffer | sponge. Once this stack is prepared, it is placed in the transfer chamber, and a current of suitable magnitude is applied for a certain period of time according to the number of transferences and membrane being used (commonly 300 mA, 25V).

Table 4.12 Western Blot buffers

TBS-Tween/Triton-Buffer		Transfer Buffer		Buffer A		TBS Buffer	
20 mM	Tris-HCl pH 7.5	25 mM	Tris-HCl	100 mM	Tris-HCl pH 9.5	10 mM	Tris-HCl pH 7.5
0.5 M	NaCl	192 mM	Glycine	100 mM	NaCl	150 mM	NaCl
0.05 % (v/v)	Tween 20	0.1 % (w/v)	SDS	5 mM	$MgCl_2$		
0.1 % (v/v)	Triton X-100	20 %	Methanol				

To obtain an adequate protein band(s) transference from the initial separating gel, the PVDF membrane needs to be activated since in its original form it is very hydrophobic and has comparatively poor binding properties. Activation is normally achieved by treatment with methanol which makes the membrane more hydrophilic allowing for aqueous solutions to penetrate it and thereby increasing the surface area accessible for binding. Typically, the membrane is treated with methanol for a short period, followed by several rounds of washing in Transfer buffer. It is essential that the membrane is not allowed to dry as it then resumes its hydrophobic, low-binding properties.

After transference, the membrane is blocked to reduce non-specific protein interactions between the membrane and the first antibody. This is achieved by placing the membrane in a 5 % milk in TBS-Tween/Triton buffer (Table 4.12).

The primary antibody, specific for the protein of interest, is then incubated with the membrane. The first antibody used for detecting Amt-1 variants, based on their C-terminal hexaHis-tag, was the Tetra-His[TM] Antibody, BSA free, Mouse monoclonal IgG_1(Qiagen), prepared in 1:20000 in TBS.

After rinsing the membrane in TBS Tween/Triton buffer to remove unbound primary antibody, a secondary antibody against the primary antibody is incubated with the membrane next. The secondary antibody is typically linked to an enzyme that allows for visual identification (alkaline phosphates-AP) of positive bands.

For detection a NBT/BCIP mixture freshly prepared was used. 5-Bromo-4-chloro-3-indolyl phosphate (BCIP) is the alkaline-phosphates substrate which reacts further after the dephosphorylation of BCIP to develop a dark-blue color as an oxidation product. Nitro blue tetrazolium chloride (NBT) serves as the oxidant and also produces a dark-blue staining. It intensifies the color development and makes the detection more sensitive. Bands corresponding to the detected protein of interest will appear as dark regions on the membrane. Band densities in different lanes can be compared providing information on relative abundance of the protein of interest [67].

4.12.17. Determination of the ammonium content in solution

Ammonium can be determined by different methods: (i) the indophenol method [68], (ii) by using enzymes such as glutamate dehydrogenase [69], carbamate kinase [70] and carbymoyl-phosphate synthetase 1 [71]. The method using the Nessler reaction [71] is recognized to be simple and fast. At this point it is necessary to note the fact that the Nessler reaction is extremely sensitive to the amount of reagent used, the incubation time and the temperature of the experiment [73]. Disadvantage is coming from its toxic effect if swallowed, inhaled or absorbed through the skin. It presents a neurological hazard and may act as a carcinogen and be a reproductive hazard. It is corrosive and causes burns.

4.2.17.1 Ammonium uptake assay

In vivo determination of ammonium uptake was carried out in *ΔamtB* C43(DE3) *E. coli* strains. When cells reached an adequate growth level and protein expression yields were satisfactory (checked in parallel by the OD_{600nm} and Western blots of samples taken at various growth times), an ammonium source (NH_4Cl) was subsequently added and the uptake of ammonium from the growth culture was measured at different times.

One colony was inoculated in 10 mL of LB medium (supplemented with 100 μg/mL ampicillin) and let grow for 8-10 hours at 37 °C. This pre-culture was harvested by centrifugation at 4 000 g, for 10 min, the cell pellet was washed in M9 salts solution, centrifuged again (4000 g, 10 min) and the resulting pellet was further resuspended in 500 mL of fresh M9 Minimal medium, supplemented with 5 mL of L-arginine (stock 10 mg/mL) as the sole nitrogen source. Cells were placed to grow overnight at 25 °C with 550 rpm stirring. At an OD_{600nm} of 0.5-0.6 protein expression was normally induced with 1 mM IPTG. After further incubation (usually for approximately 3-4 hours), and when an OD_{600nm} of 0.8-0.9 was reached, ammonium chloride was added to a final concentration of 1 mM. At regular intervals, 1 mL aliquots of the culture were taken and centrifuged. The supernatant was then used immediately for assaying ammonium with the Nessler reaction.

4.12.17.1.1. Nessler reaction

Nessler's reagent (Fluka Analytical) is a 0.09 mol/L solution of potassium tetraiodomercurate (II) ($K_2[HgI_4]$) in 2.5 mol/L potassium hydroxide. In the presence of ammonium, a yellow color develops (see reaction down) (at higher concentrations, a brown precipitate may form) and its intensity, proportional to the amount of ammonia in solution, can be measured at 475 nm [74].

$$NH_4^+ + 2[HgI_4]^{2-} + 4OH^- \rightarrow HgO \cdot Hg(NH_2)I + 7I^- + 3H_2O$$

To make ammonium quantifications in the above described growth medium solutions, 100 μL of the medium supernatant was pipetted into 1.5 mL tubes and 1 mL of Nessler reagent was added. After an incubation step carried at room temperature for 20 min to allow for colour development, the absorption was measured at 475 nm. Ammonium concentration was estimated after a calibration curve (done following the same procedures).

4.12.18. Protein crystallization

Proteins, like any other molecule, can be prompted to form crystals when placed in the appropriate conditions. In order to crystallize a protein, the purified protein undergoes slow precipitation from an aqueous solution. An optimal crystal (good diffracting properties) should have a regular repeating three-dimensional arrangement of the protein or entities that it makes it. This repetitive unit is called the unit cell, which is the smallest region of the crystal that represents the whole crystal. This regular arrangement of unit cells, which can be defined by lengths and angles, form the crystal lattice. Each protein can be crystallised in a different three-dimensional array, or crystal form, belonging to one of 65 different chiral space groups [75].

For a better understanding of the crystallization process phase diagram is rather helpful (Figure 4.3). In the phase diagram, three stages are involved: nucleation, crystal growth and cessation of growth. When the protein concentration reaches the nucleation region, nuclei are formed. The protein concentration decreases as nuclei form, and the metastable phase is reached in the phase diagram. In this region, it is likely that crystal growth from the nuclei occurs and continue to grow, but no additional nucleation takes place. Cessation of crystal growth is thought to occur when the molecules reach exchange equilibrium between the solution and crystal phases and/or the concentration is depleted such that the undersaturated region is reached.

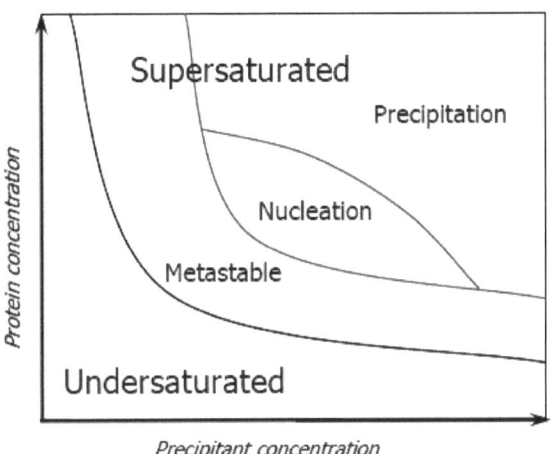

Figure 4.3: Phase diagram for protein crystallisation: Precipitation occurs when the protein concentration is to high and the protein molecules aggregate. Nucleation is the spontaneous aggregation of nuclei, but no further growth occurs. Metastable stage is when growth of existing nuclei and crystals occurs but not new nucleation. The undersaturated state means that the protein molecules remain in solution. Picture taken from reference [76].

The following crystal quality parameters are important for a successful X-ray crystallographic measurement: (i) the crystal should have an adequate size (typically larger than 0.1 mm in all dimensions), (ii) crystal has to be in a cryoprotective solution (frozen water molecules themselves give rise to diffraction which interferes with the diffraction pattern from the crystal or in the worst case destroy the protein crystal. The addition of cryo-protectants to the drop solution leads to vitrification of the crystal drop, that is then free from any other crystalline structure besides the protein) and, (iii) the protein should have an homogeneous composition to allow for a regular crystal formation without packing failures [77].

Crystallization experiments were performed with the protein concentrated to 10 mg/mL, by vapor diffusion method in sitting drop crystallization plates. 1 μL of protein solution was mixed with 1 μL of precipitant solution and this mixture was equilibrated against 200 μL of reservoir solution. The whole setup was sealed with crystal clear tape and stored at 20 °C. Crystallization plates were observed daily.

4.12.18.1. Crystals testing

The quality of a crystals is typically assessed by exposing them to X-ray radiation [78]. Only specific wavelength X-rays in the range of 0.7 to 2.5 Å are used in protein crystallography. Because the wavelength of X-rays is comparable to the interatomic distances, they are ideally suited for probing the structural arrangement of atoms in the crystal. Radiation damage can be minimized by cooling the crystals to 100 K during the measurement. Diffraction experiments were carried out at the Swiss Light Source (SLS). In an X-ray diffraction measurement, a crystal is mounted on a loop that is then placed on a goniometer and rotated through a small angle (typically 0.5-1°) during exposure to X-rays. When a crystal is placed in the path of an X-ray beam its atoms act, owing to the forced vibrations of the electrons, as secondary sources emitting X-rays to each direction. The frequency and wavelength of these emitted rays are identical to those of the incident beam. Because a crystal is constructed of atoms or molecules arranged in a regular spatial pattern, only in certain directions the individual scattered wavelets may recombine in phase to produce a strong reinforced but deviated beam. This produces a diffraction pattern of regularly spaced spots known as reflections.

4.12.18.2. Data processing

The processing of the diffraction data is mathematically complex and are part of the well established algorithms in many software packages and program suites. The first step in the processing involves the determination of the crystal system and of the unit cell dimensions as accurately as possible. In addition, at this stage we determine the orientation of the crystal in the beam [79]. When the unit cell dimensions and crystal orientation are known, indexing can be carried out [80]. In this, each spot on the image is assigned an index, quoted as three integers: h, k, and l. Computer programs for autoindexing do this by calculating a prediction of what the diffraction image will look like from the cell dimensions and orientation, then attempting to fit the real image with the predicted one. The next stage of the data processing is the measurement of the intensities of the spots. Protein crystals diffract weakly because they are composed mainly of light atoms and they have large unit cells. The larger the volume of the protein crystal, the stronger its diffraction. Intensities of diffracted spots vary as a result of both the amplitude of the diffracted waves and their phase relation. These factors cannot be deconvoluted at this stage; accordingly, the accuracy of the measurement of the intensities is of paramount importance. One of the programs used most frequently in protein crystallography which do both autoindexing and intensity measurements is MOSFLM [81]. A scale factor must be allocated so that the intensities of all the images in the data set can be related. The first image is usually allocated a scale factor of 1 and all the subsequent images will be scaled up to this. The output from scaling is a file that contains the index of each spot and its measured intensity. This file must be sorted so that the spots are listed in numerical order according to their index.

4.12.18.3. Phase problem

In order to solve the three-dimension structure of a protein, besides the information collected directly from the diffraction images regarding spot intensities and relative position in space to each other, one needs to solve the phase problem. Namely, during a standard protein crystallographic experiment only the intensities of the diffracted X-ray beams are recorded, from which the amplitudes can be obtained. Unfortunately, the relative phases of these wavelets, crucial for reconstructing the image of the molecule, are lost. For protein structure determination we can do it in one of several ways. Solving this ''phase problem'' is the central problem in crystallography. Many methods have been developed to deduce the phases for the reflections including isomorphous replacement, statistically based direct methods, molecular replacement method, etc [82, 83].

1. <u>Molecular replacement</u>

If the structure of a similar protein is available in the data base, then it might be possible to solve the new structure by a process called molecular replacement [84]. This process involves using the previously known model to rotate and translate into our new crystal system until a good match to our experimental data is reached and a possible solution is obtained. If the process is successful, then one can proceed to calculate the phases and produce an electron density map. This is done using Fourier transformations [85, 86]. The model used to solve the phase problem in the various *Af*Amt-1 variants was the wild-type *Af*Amt-1 (PDB-ID 2B2H). Molecular replacement was made using the program MOLREP [87] from the ccp4 suite.

2. Isomorphous Replacement methods

If there is no enough similar model, then we can use isomorphous replacement methods [88] which includes introducing one or more heavy atoms into specific sites within the unit cell without perturbing the crystal lattice. Heavy atoms are electron dense and give rise to measurable differences in the intensities of the spots in the diffraction pattern. There are two ways for introduction heavy atoms: (i) soaking the crystal in a heavy atom-containing solution, or (ii) co-crystallization (growing the crystals in the presence of a heavy atom). By measuring the resulting differences in each reflection, it is possible to derive some estimate of the phase angle using vector summation methods.

4.12.18.4. Model refinement

Once we have an initial model we need to refine it against our experimental data. This, hopefully, will have the effect of improving the phases resulting in clearer maps and therefore better models. Refinement was done using REFMAC [89], a program from the CCP4i software suite. Typically several rounds of refinement and model building cycles are done several times until little or no further improvements are seen. At such stage we expect a crystallographic R-factor [90] below 25%. This is a measure of the agreement between the model and the experimental data – the lower the R-factor, the better the model is. After refinement, the model quality should be evaluated by electron density, the crystallographic residual, geometry parameters and biological sense, etc. In principle, obtaining a high quality of a refined model is only possible if high quality data are available, that is: a high degree of completeness, high redundancy, high I/σ(I) ratio, high resolution, etc. High quality data are also extremely important for successful substructure determination and the subsequent phasing process.

Once the model of a molecule's structure has been finalized it is often deposited in a crystallographic database such as the Protein Data Bank (www.pdb.org).

5. Results and Discussion

5.1. Site-directed mutagenesis of *Af*-Amt1

In order to explore the validation of the recruitment site, the possibility of ammonium deprotonation and reprotonation events, or the role of the C-terminus in transport control, among others we set up to constructed different mutants of the Amt-1 (Table 5.1).

As stated in the Introduction, the remarkable conservation of two pore histidines (H157 and H305) led to the hypothesis that may play a key role in allowing the substrate to cross the central part of the channel [16]. They might serve as proton acceptors for entering ammonium (NH_4^+) that could then traverse the central part of the channel as ammonia (NH_3) and become reprotonated on the other side. Alternatively, efficient substrate conductance might require such a narrow and mainly hydrophobic pore with a few rather precisely oriented hydrogen bond acceptor or donor functions for weak, stabilizing interactions with water/ammonia that still permit rapid diffusion [91].

A258 is in helix VIII close to H157 (about 4.3 Å away) and our starting hypothesis to validate is that it may help H157 to be properly positioned. V156 is direct the neighbour of H157 (in the same helix V) and it may also promote for a proper positioning of this histidine residue.

G309 is located in the helix X together with H305. Its identity properties could again change the helix positioning and consequently the positioning of H305.

W201 is very interesting because it seems to be in a crucial position as well. It lies in helix VI together with F204 (right bellow, about 3.5 Å away) and may have a still unassigned, important role in the substrate transport.

D160 from AmtB has been shown to be required for transport function and is proposed to participate in ammonium binding [92]. The homologous aspartate residue Amt-1 is D149 is located on top of helix V (with H157). A possible structural and/or functional role for this residue that seems to hold helix V in the right position and have hydrogen bonds to W137, at the same time.

The role of residues N205 and T261 are still unclear. N205L and T261V mutations are done after Matthias Ullman´s molecular dynamic calculations. Simulations studies pointed to these two residues positioned open interesting question about possible place for protonation/deprotonation.

For future experiments it should be interested to test whether the C-terminus of Amt-1 is required for transport function. In this context different point mutations and deletions are going to be introduce into the C-terminus.

Each mutation was done according to procedures described in Materials and Methods and were confirmed by sequencing (GATC Biotech). The results were analyzed using Chromas (Technelysium), a program to display the chromatogram files from ABI automated DNA sequencers, together with Clustal W2 [93] to do sequence alignments. Looking at the chromatogram with Chromas is often essential to verify the presence of missing bases or wrongly automatic base anotations.

Appendix 7.2 shows the DNA sequence alignments for the wild type *amt-1* and its variants, obtained using T7 forward and reverse primers.

Table 5.1: List of mutants which sequences were confirmed: (i) in previously work, (ii) in this work, and (iii) variants for which sequence still need confirmation.

Previous confirmed mutants	Mutants confirmed in this work	Mutants which still need confirmation
D149E, D149L, D149N	D149A	A279G
A279G	N205L, T261V	W201G
H157F, H157N, H157D, H157F/H305F, H305F	I304E	H157F/I304E
Δ156	W137L	H157A/H305F
G309S	W201A, W201F	H157F/H305A
A258S, A258T, A258V	H157E	H157D/H305F
W137H	H305E	H157F/H305D
V156L, V156A, V156S	A112L	H157A/H305A
		H157E/I304E
		H305E/I304E
		H157F/I304E

5.2. Heterologous overexpression in *Escherichia coli* C43 (DE3) competent cells

In order to optimize the expression level of Amt-1 variants in *E.coli* C43 (DE3) cells test expression were performed. Small scale (0.5 L) cell culture growths were monitored by measuring the optical density at 600nm (OD_{600nm}) in aliquots of 1 mL, taken at chosen growth time points. Induction was initiated by adding 1 mM IPTG at OD_{600nm} 0.6-0-8 (Table 5.2).

Table 5.2: Optical density (OD$_{600nm}$) records for the test expression cultures of A112L, H157E, W201F, D149A, H305E, I304E, W201A, D149E, respectively. t_0- time of induction with 1mM of IPTG; t_1-, t_2-, t_3- one, two and three hours after induction with 1mM of IPTG, respectively.

A112L		H157E		W201F		D149A	
OD$_{600nm}$	Time (min)	OD$_{600nm}$	Time (min)	OD$_{600nm}$	Time (min)	OD$_{600nm}$	Time (min)
0.045	0	0.077	0	0.079	0	0.145	0
0.353	123	0.375	143	0.356	156	0.456	172
0.559	176-t_0	0.745	165-t_0	0.597	189-t_0	0.754	194-t_0
0.976	236-t_1	1.345	234-t_1	0.861	249-t_1	1.145	224-t_1
1.234	296-t_2	1.534	294-t_2	1.334	309-t_2	1.156	284-t_2
1.345	356-t_3	1.563	354-t_3	1.761	369-t_3	1.221	344-t_3

H305E		I304E		W201A		D149E	
OD$_{600nm}$	Time (min)	OD$_{600nm}$	Time (min)	OD$_{600nm}$	Time (min)	OD$_{600nm}$	Time (min)
0.157	0	0.087	0	0.059	0	0.067	0
0.354	145	0.275	123	0.463	96	0.321	110
0.81	175-t_0	0.731-t_0	167-t_0	0.606-t_0	131-t_0	0.721	156-t_0
1.163	235-t_1	0.937-t_1	227-t_1	0.997-t_1	191-t_1	1.056	216-t_1
1.283	295-t_2	1.134-t_2	287-t_2	1.084-t_2	251-t_2	1.276	276-t_2
1.53	355-t_3	1.144-t_3	347-t_3	1.262-t_3	311-t_3	1.345	336-t_3

The OD$_{600nm}$ values may be plotted against time to generate a growth curve. All growth curves (Figure 5.1) show slightly exponential behaviour without significant lag-phase in due to the over-night cell preincubation. The growth (grown at 37 °C) in log phase showed exponential growth from the moment of inoculation freshly LB medium with the overnight pre-culture and declined to the stationary phase between two and three hours after induction with 1 mM IPTG. The exponential phase in case of D149A (grown at 30 °C) showed an increasing of OD$_{600nm}$ during the first three hours after inoculation with the overnight preculture and continued growing one hour after induction with 1 mM of IPTG. The stationary phase has been reached at an OD$_{600nm}$ of around 1.145. It should be mentioned that the exponential phase of cell growth at 37 °C was reached faster than the one at 30 °C. This is not surprising, because the higher the temperature in growth of the *E. coli* promotes faster metabolism [94].

Figure 5.1: Growth curves of the test expressed Amt-1 variants N205L, H157E, W201F, D149A, H305E, and T261V, respectively. t_0-time of induction with 1mM of IPTG; t_1-, t_2-, t_3- one, two and three hours after induction with 1mM of IPTG, respectively.

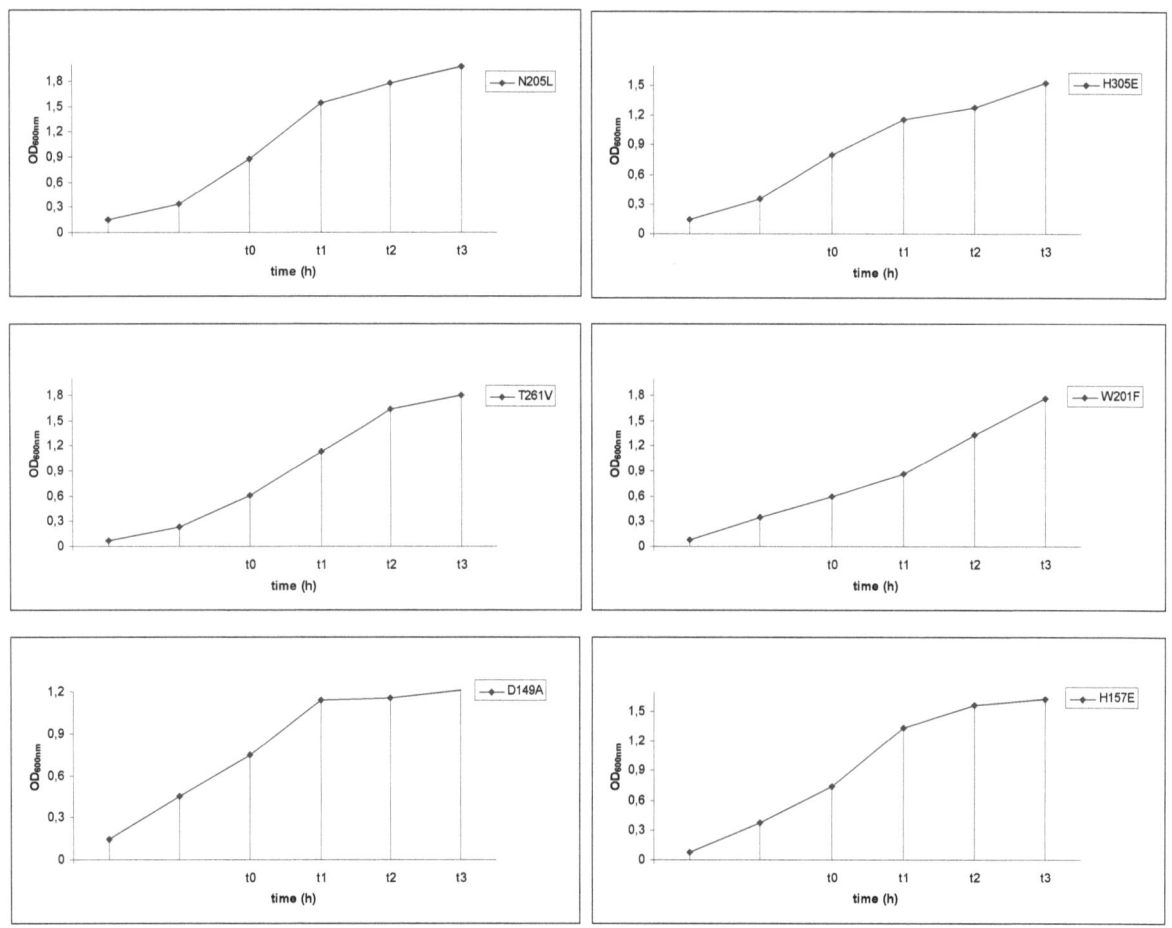

The levels of protein expression were monitored by Western Blot. Samples were collected right before induction (t_0), one hour (t_1), two hours (t_2), and three hours after induction (t_3) when the stationary phase was reached.

At t_0, when IPTG is not present in the medium, the lac repressor binds very tightly to the lac operator (present in the pET-21a vector) and thus the lac operon is blocked and there is not any significant expression of protein. At t_1, one hour after 1 mM of IPTG was added to the medium, it binds to the lac repressor causing a change in its shape. Thus altered, the repressor is unable to bind to the lac operator, allowing the T7 RNA polymerase to proceed transcription of the lac genes and the encoded protein. The expression of protein increases rapidly as the amount of mRNA transcribed from gene increases such that within a few hours the expressed protein is one of the most prevalent components of the cell. One of the most important features of the pET expression system involves the fact that the target gene is not transcribed unless the T7 RNA polymerase is present. One advantage of IPTG for *in vivo* studies is that since it can not be metabolized by *E. coli* its concentration remains constant and the rate of expression of *lac p/o*-controlled genes, is not a variable in the experiment. IPTG intake is dependent on the action of lactose permease [95].

Figure 5.2 shows examples of Western blots profiles for the overexpression of N205L, T261V and D149A, respectively.

Taken together, the results from the Western blot and growth curves for D149A shows almost the same level of protein expression after one and two hours upon IPTG induction which is in agreement with the growth curve rapidly increasing of OD_{600nm} during one hour and then reach stationary phases within two hours with slightly increasing of turbidity between two and three hours. T261V and N205L blots show an increase in the protein expression during three hours with possible increasing of protein expression after in according to the growth curve which did not reach stationary phase after three hours.

Figure 5.2: Western blot of test expression cultures N205L, T261V and D149A, respectively, using tetra-His antibody against C-terminal hexa-His-tag. M-protein marker, t_0-time of induction with 1mM of IPTG, t_1-, t_2, t_3- one, two and three hours after induction with 1mM of IPTG, respectively.

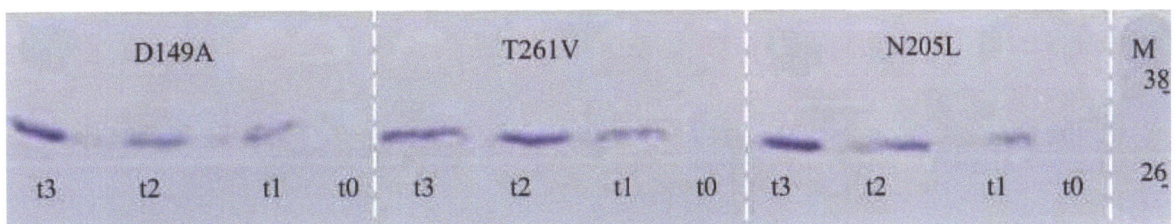

5.3. Protein Purifications and Crystallization

The purification protocol followed the guidelines that were previously established for the Amt-1 purification [96]. Figure 5.3 and 5.4 show the purification chromatograms of T261V and N205L, respectively. In both cases, purification was initiated with approximately the same amount of cells (24.4 g for T261V and 25.3 g for N205L). To decrease the interaction of protein impurities to the column material further purifications were tested with different amounts of imidazole in Buffer A (5 mM, 10 mM, 15 mM, 20 mM, 25 mM 30 mM imidazole). This, however, did not add to the final protein yields or purity levels. Purification of T261V and N205L resulted in smaller elution peaks (peak 2) from the HiTrap^{TM} FF affinity column in comparing to results obtained with wild-type Amt-1 elution peak (200 mAbs for both variants and 2000 mAbs for the wild-type, respectively).

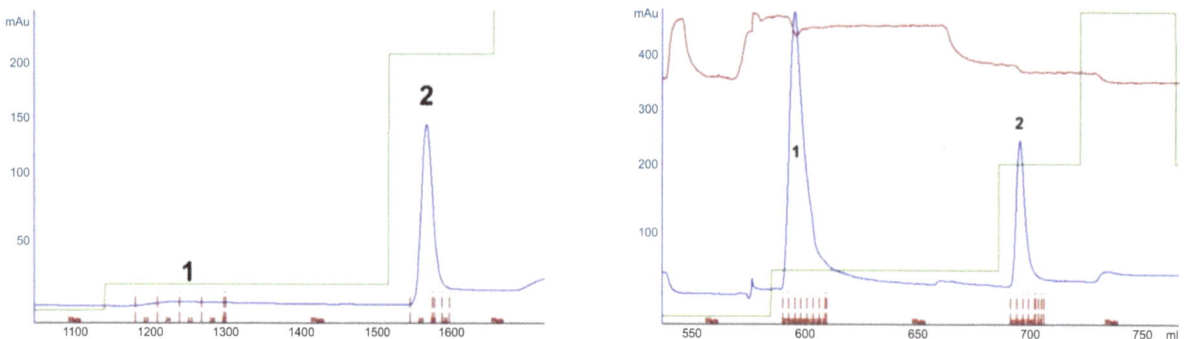

Figure 5.3: Chromatogram of the HiTrap^{TM} FF affinity column of the T261V mutant (left) and N205L (right) after detergent exchange, from 0.03% D$_{12}$DM to 0.05% LDAO. A washing step (peak 1) has done at 17% of Buffer B (containing 85mM imidazole) and the elution step (peak 2) has done with 50% of Buffer B (containing 250 mM imidazole). The absorption profile, measured in mAU at 280 nm, is shown in blue. The red line shows the conductivity and the green line shows the percentage of Buffer B along the course of the chromatography.

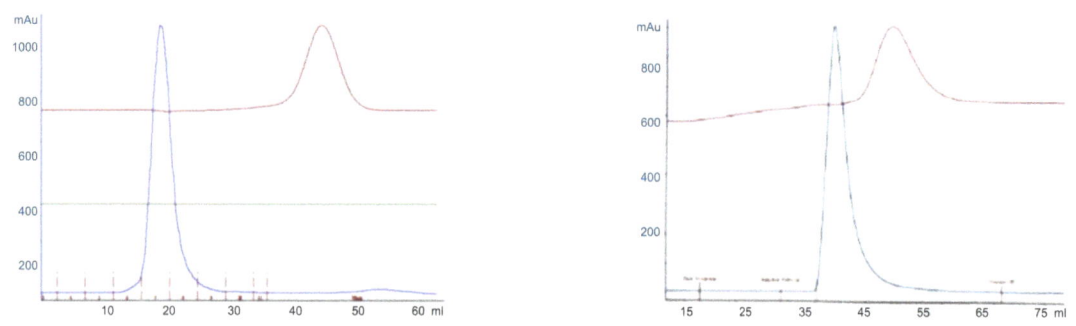

Figure 5.4: Chromatogram of the HiTrap^{TM} desalting column with Sephadex™ G-25 Superfine Resin of the T261V mutant (left) and the N205L mutant (right) with Buffer C containing 20 mM Tris-HCl pH 8.0, 100 mM NaCl, 10% glycerol, 0.05% LDAO.

The purification process and the purity level of the protein was controlled by SDS-PAGE (Figure 5.5). The gel shows two bands corresponding to the dimer (60 kDa) and the monomer (33 kDa) [97]. T261V was obtained with higher purity than N205L as can be seen from the additional bands that are visible in the SDS-PAGE for the N205L protein. These contaminations could eventually be eliminated using an additional washing step between 17 % and 50 % of Buffer B or by gel filtration chromatography (not tried).

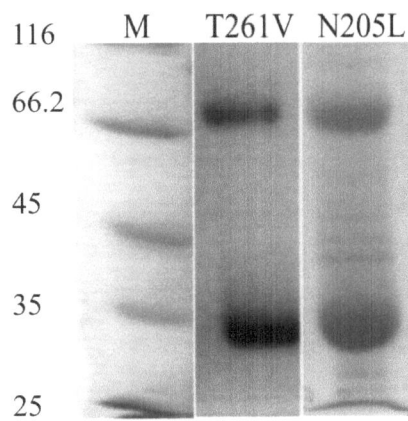

Figure 5.5: 12.5 % SDS-PAGE of purified T261V and N205L after Ni-affinity chromatography stained with Commasie Briliant Blue. M-protein marker (kDa).

All proteins were solubilized in $D_{12}DM$ but unfortunately protein purified in this detergent is not ideal for crystallization, due to this detergent long alkyl chains. For this reason, a detergent exchange step is necessary. It is known, that membrane proteins are best solubilized using long chain detergents, because of the stronger interactions due to their long hydrophobic part. However, often these detergents can inhibit totally or partially a proper crystal formation. On the contrary, too short alkyl chains may not be enough to stabilise the protein, leading to denaturation [97]. This behaviour can be understood when looking at the two main categories of membrane protein crystals on the molecular level. As stated in reference 97 "Crystals of type I consist of stacked protein molecule layers, in which hydrophobic interactions in one plane and polar protein/protein contacts between the several layers hold the crystals together. The crystals of type II do not show the hydrophobic contacts. They only show lattice contacts within the polar regions of the membrane proteins. That is why detergents with too long alkyl chains can inhibit a protein packing in crystallization process, if their steric influence prevents the adequate protein/protein contacts. To obtain membrane protein crystals, it is necessary to experiment a detergent, which allows these kind of lattice contacts, because of its chemical properties, so that the proteins can be arranged in the crystal. Based on the previous experience in former crystallization trials with Amt-1, LDAO was chosen to proceed into crystallisation experiments. Detergents with ionic or zwitterionic head groups have the advantage, that they can inhibit protein aggregations, because of the ionostatic repulsions but this strong interactions could lead to denaturation''. To choose an adequate detergent for crystallization is always a compromise, which has to be experimented.

Crystals were obtained using a reservoir solution composed of 30 % PEG 400, 100 mM Na-citrate pH 5.5 and 100 mM NaCl at a protein concentration of 10 mg/mL and incubated at 20 °C (Figure 5.6). 1 µL of protein solution was mixed with 1 µL of precipitant solution and this drop was equilibrated against 200 µL of reservoir solution. Cubic crystals usually appeared after 4-5 hours and grew for further 7 days. After the hit in this condition fine screens for N205L and T261V were established to optimise the size and diffraction quality of the crystals. By varying the salt concentration (0-300 mM NaCl), the buffer concentration (100-350 mM Na-Citrate pH 5.5), the precipitant concentration (25-40 % PEG 400), and protein concentration (10-20 mg/mL) bigger cubic crystals were found. The optimal condition for N205L was reached with 35 % PEG 400, 0.1 M NaCl, 0.1 M Na-Citrate pH 5.5, at a concentration of 10 mg/mL protein. For the T261V variant, the optimal condition was with 35 % PEG 400, 0.1 M NaCl, 0.1 M Na-citrate pH 5.5 and 15 mg/mL of protein. Flash-freezing of the crystals before data collection was done directly from the drop because the reservoir solution was already cryoprotectant.

D149A crystals appeared in a condition composed of 30 % PEG 400, 100 mM Na-citrate pH 5.5, 100 mM NaCl and 100 mM Li_2SO_4 (Figure 5.6). Small crystals appeared after one week and grew for further three weeks. Fine screening did not help in optimizing the size of the crystals. Variations of the drop size also did not produce better results. This finding could indicate that mutating D to A at position 149 can lead to drastic changes in the tertiary structure and/or to an eventual protein instability which in turn makes it difficult to crystallize.

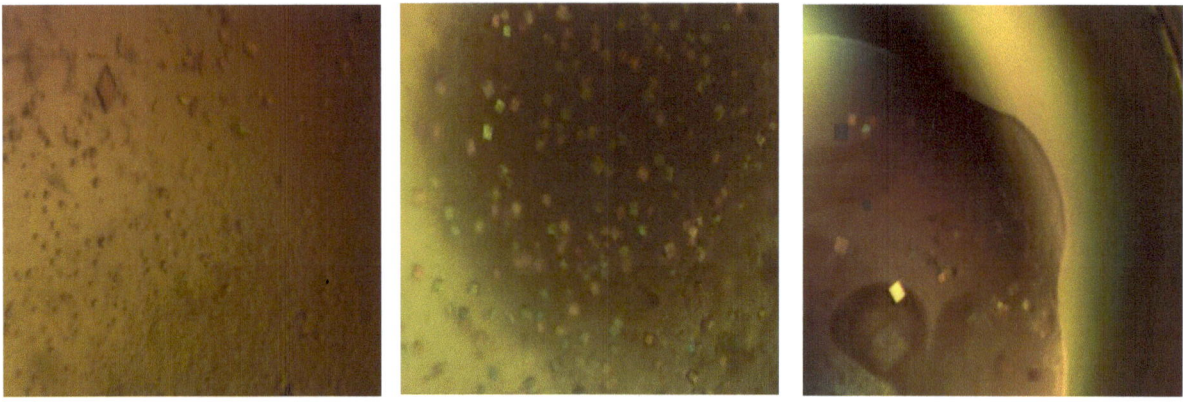

Figure 5.6: Crystals of N205L (left), T261V (middle), and D149A (right) obtained by sitting-drop vapour diffusion at 20 °C. Drop size was 1 µL mixed with 1 µL of 200 µL of reservoir solution. The reservoir solution proved to be a usable cryo condition for carrying out diffraction experiments. Crystals were flash-cooled directly in liquid nitrogen. Pictures were taken under polarized light.

5.4. X-ray Structures of Variants N205L and T261V

Two of the functionally characterized Amt-1 variants were crystallized in almost identical conditions as the wild-type Amt-1 protein. This allowed solving their crystal structure at a resolution of 2.2 Å for T261V and 2.7 Å for N205L (Table 5.3 for data collection and refinement statistics). The structure of these variants was solved by molecular replacement with programs from the CCP4 suite [87], using wild type Amt-1 (PDB-ID 2B2H). For refinement, 5 % of all reflections were chosen at random and used as a test set for cross-validation [98]. Subsequent steps of model building were carried out in COOT [99], and refined using REFMAC.

As mentioned in chapter 5.1, those two residues are proposed to be a place where deprotonation/protonation even could occur, based on molecular dynamic simulation studies.

When comparing the structure of both N205L and T261V mutants with wild-type Amt-1, structural changes were restricted to the mutated residues (Figure 5.7 and 5.8). Exceptions go for a clear observable change in the conformation of F204 and a minor change in the conformation of the F96 in the case of the N205L variant. A detailed inspection of the T261 structure also does not show big conformational changes, neither in the overall protein model, nor in the neighboring residues, with the exception of a minor change in residue F96.

Table 5.3: Data collection and refinement statistics

	N205L	T261V
Resolution (Å)	2.75	2.2
Space group	H3	H32
Unit cell (a, b, c)	a=b=110.38 c=321.97	a=b=111.405 c=136.908
Resolution range (Å)	55.19-2.7	55.83-2.2
Number of unique reflections	39151	16799
Number of total reflections	113647	165826
Number of atoms in model	2895	3006
Completeness (%)	97.51	99.62
Multiplicity	2.9	9.9
Mean I/σ	4.0	7.6
Rmerge (%)	17.4	24.5
Rsym (%)	20.41	29.47
Rfree (%)	24.69	32.48
Average B value		
Protein main chain	40.571	17.703
Protein side chain	41.818	17.308
Protein all atoms	41.144	17.521
Cruickshanks DPI	0.3573	0.5371
r.m.s.d in bond lengths (Å)	0.024	0.031
r.m.s.d. in bond angles (°)	2.0434	2.451

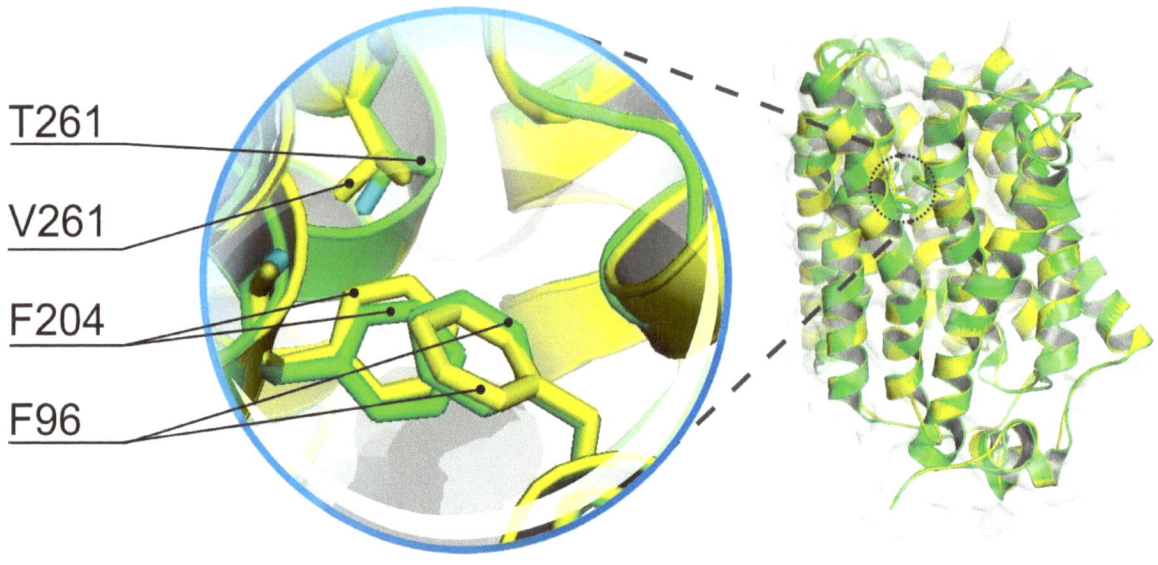

Figure 5.7: Superposition of the wild type Amt-1 structure (green) with T261V variant structure (yellow).
The right picture shows the monomer in cartoon representation with slightly surface representation in the background. The cytosolic part is located below and the extracellular part is above. Enlraged part on the left side represents mutated residue V261 (yellow) and wild-type residue T261 (green) with minor structure changes in the surrounding pheynilalanine residues (F204 and F96, respectively).

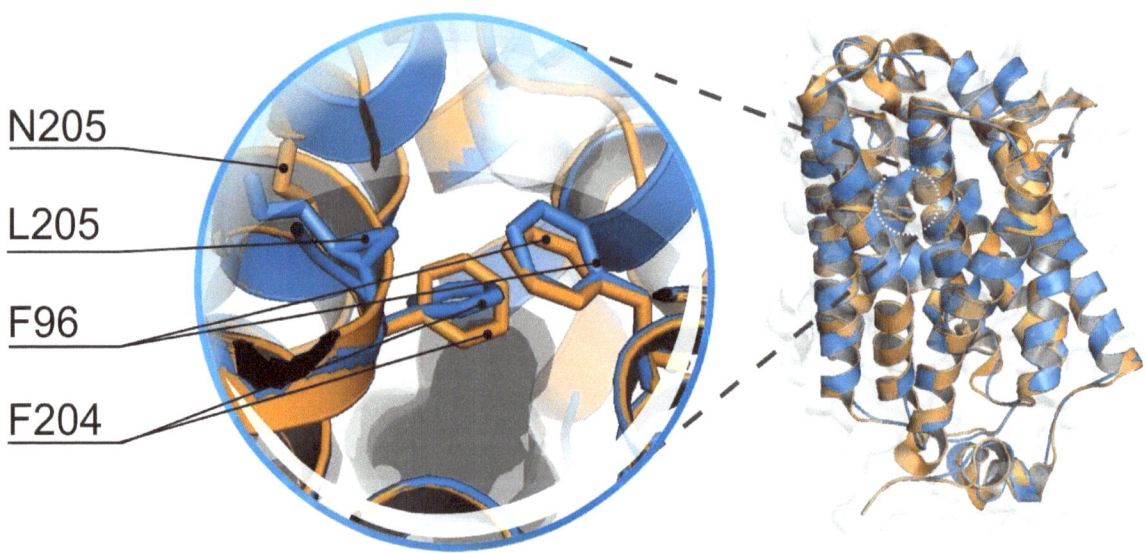

Figure 5.8: Superposition of the wild type Amt-1 (orange) with N205L variant structure (blue).
The right picture shows the monomer in cartoon representation with slightly surface representation in the background. The cytosolic part is located below and the extracellular part is above. Enlraged part on the left side represents mutated residue L205 (blue) and wild-type residue N205 (orange) with minor structure changes in the surrounding pheynilalanine residues (F204 and F96, respectively).

5.5. Ammonium uptake assay

Ammonium uptake assays were carried out in BL(21)C43 *E. coli* cells that were knocked-out in their *amtB* gene. Following transformation with a plasmid carrying the *A. fulgidus amt* gene of interest they were set to grow on minimal medium supplemented with arginine as the sole nitrogen source. One hour after induction with IPTG, ammonium was added to the growth medium and measured (475 nm) from the culture supernatant at regular intervals using the Nessler reagent. For each experiment a negative control was prepared by transforming *ΔamtB*C43(DE3) strains with an empty pET-21a vector following identical steps for ammonium uptake assay as those taken for the correspondent variants.

The calibration curve (Figure 5.9) was treated with a linear regression fit to the experimental points consisting of the measured A_{475nm} at known concentration of ammonium chloride (0.05-1 mM). In this context we should mention the concept of the correlation coefficient R^2, a statistical measure of how well the regression line approximates to the real data points. An R^2 of 1.0 indicates that the regression line perfectly fits the data. [100].

Figure 5.9: Standard curve for the Nessler reaction. The graphic shown the experimental results and linear regression fitting (R^2=0.9973).

NH$_4$Cl (mM)	0.05	0.05	0.05	0.1	0.1	0.1	0.2	0.2	0.2	0.5	0.5	0.5	1	1	1
A$_{475nm}$	0.034	0.031	0.029	0.054	0.053	0.049	0.089	0.083	0.081	0.221	0.215	0.219	0.467	0.459	0.466

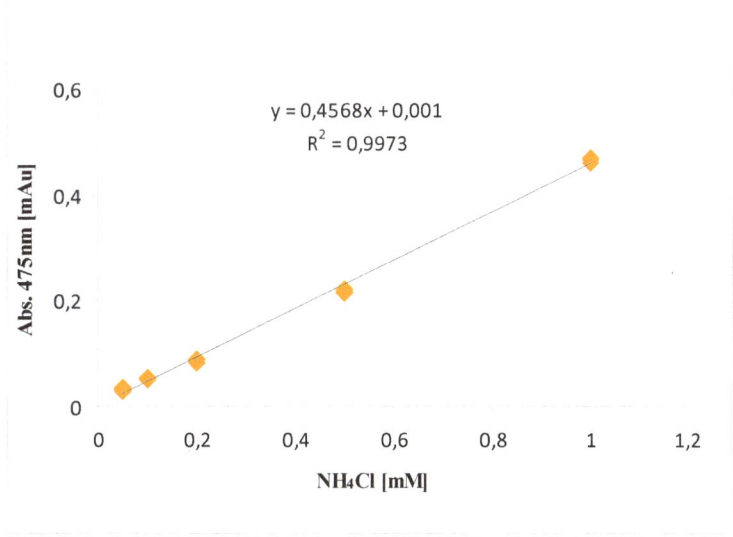

For each mutant, three colonies from three independent tansformationns were picked out from LB agar petri dishes and each one was inoculated in 0.5 L of LB medium for an overnight pre-culture. Triplicates of 1 mL from each colony growth were taken for the Nessler reaction, during their growth in minimal medium. Each experiment was repeated two times, giving in the end 18 data points for each variant. The Nessler assay was done immediately after the samples were taken because earlier experiments showed that saving aliquots either at 4 °C or -20 °C dramatically changes the final records.

As stated in the Material and Methods, the concentration of ammonium was monitored in strains that lack *amtB*. In these conditions, the basal decrease of ammonium concentration in the medium was only 1-2 %. These results indicate that there is no other transporter/channel that effectively takes up ammonium under these conditions. In this study, the function of Amt-1 was assessed by analysing some of its variants.

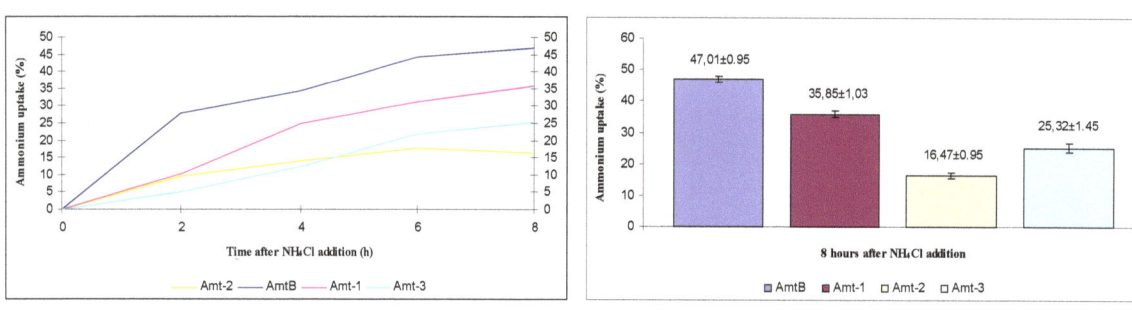

Figure 5.10: *In vivo* **determination of ammonium uptake in** *ΔamtB* **C43(DE3)** *E. coli* **strains.** Comparison of the *in vivo* ammonium uptake capacity between plasmids carrying the *A. fulgidus amt* gene (*amt-1 -2, -3*, respectively), and plasmid carrying the *E. coli amtB,* following transformation in BL(21)C43 *E.coli* cells that were knocked-out in their *amtB* gene. The graphic on the left shown ammonium uptake within 8 hours after addition of 1 mM NH_4Cl into minimal medium. The graphic on the right shown in the histogram representation the ammonium uptake at a chosen time point (8 hours) after NH_4Cl addition. Values are means ± S.D (n=18).

Figure 5.10 shows a comparison of the ammonium uptake between all three *A. fulgidus* Amt proteins and AmtB from *E. coli*. The reason for the highest maximum of ammonium uptake obtained for AmtB (47.01 % ± 0.95) could simply reflect the assay temperature. In this assay the growth temperature was 37 °C, an optimal temperature for *E. coli* [101] but far too low for the hyperthermophilic *A. fulgidus* Amt proteins (83 °C) [31]. The lowest activity recorded for *A. fulgidus* Amt-2 (16.47 % ± 0.95) might be linked to its lower sequence homology to Amt-1 and Amt-3 (Figure 3.4).

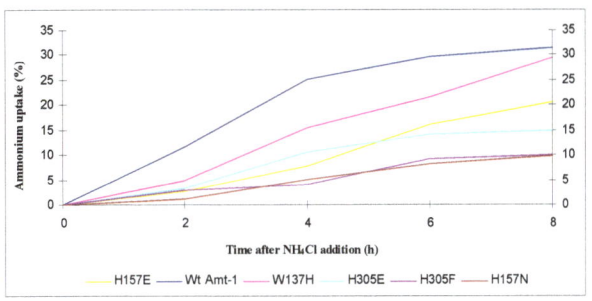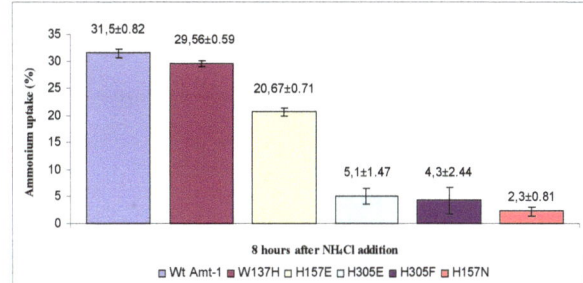

Figure 5.11: *In vivo* **determination of ammonium uptake in** *ΔamtB* **C43(DE3)** *E. coli* **strains.** Comparison of the *in vivo* ammonium uptake capacity between plasmid carrying the wild-type *A. fulgidus amt-1* gene and plasmids carrying different variants of *amt-1* (H137E, W137H, H305E, H305F, H157N, respectively) following transformation in BL(21)C43 *E.coli* cells that were knocked-out in their *amtB* gene. The graphic on the left shown ammonium uptake within 8 hours after addition of 1 mM NH_4Cl into minimal medium. The graphic on the right shown in the histogram representation of ammonium uptake at a chosen point time (8 hours) after NH_4Cl addition. Values are means ± S.D (n=12).

Figure 5.11 show that both plasmids carrying the *A. fulgidus amt-*1 (31.5 % ± 0.82) and W137H (29.56 ± 0.59) effectively remove approximately 30 % of ammonium from the medium after eight hours. Except for the partially active H157E mutant (20.67 % ± 0.71), neither H157N (2.3 % ± 0.81), H305F (4.3 % ± 2.44) nor H305E (5.1 % ± 1.47) were significantly active. These findings demonstrate that these two histidines are not functionally equivalent. It seems that the activity of Amt proteins depends on a combination of specific spatial and chemical properties [91]. But at this point these results should be considered as a preliminary and they demand additional repeating and optimization especially in the protein purification in order to get more accurate informations. In this case, the Western blot representation (Figure 5.12) is not good a indicator of estimating of protein expression level.

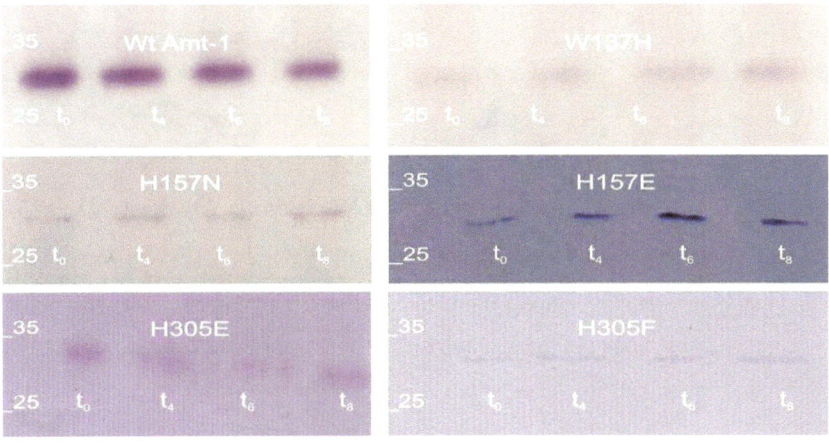

Figure 5.12: Amt-1 variant expression in *ΔamtB* C43(DE3) *E. coli* strain using tetra-His antibody against C-terminus hexa-His-tag. M-protein marker; t_0-, t_4-, t_6-, t_8- zero, four, six and eight hours after adding 1 mM of NH_4Cl into the culture medium, respectively.

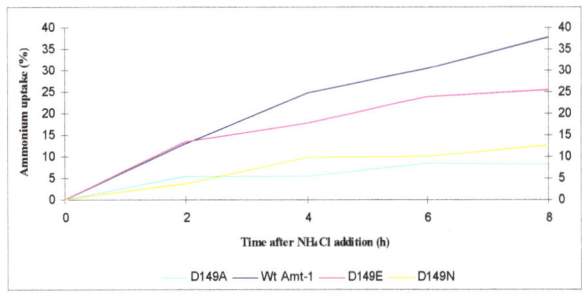 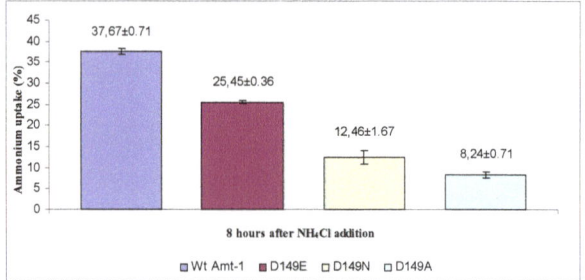

Figure 5.13: *In vivo* determination of ammonium uptake in *ΔamtB* C43(DE3) *E. coli* strains. Comparison of the *in vivo* ammonium uptake capacity between plasmid carrying the wild-type *A. fulgidus amt-1* gene and plasmids carrying different variants of *amt-1* (D149A, D149E, D149N, respectively) following transformation in BL(21)C43 *E.coli* cells that were knocked-out in their *amtB* gene. The graphic on the left shown ammonium uptake within 8 hours after addition of 1 mM NH_4Cl into minimal medium. The graphic on the right shown in the histogram representation of ammonium uptake at a chosen point time (8 hours) after NH_4Cl addition. Values are means ± S.D (n=12).

Figure 5.13 shows different ammonium capacities for a series of D149 mutants. As depicted in this Figure, plasmid carrying the *A. fulgidus amt*-1 removed around 40 % (37.67 % ± 0.71) of ammonium from the medium within eight hours, while D149E removed only 25% (25.45 % ± 0.36) ammonium during the first six hours and kept this level constant untill 8 h. Replacement of aspartate with its amide derivative, asparagine resulted in decreasing of ammonium uptake capacity (12.46 % ± 1.67). Substitution of aspartate into a smaller and neutral residue such as alanine led to an almost loss of transport function (8.24 % ± 0.71). To check whether this decrease of transport function is due to the instability of the protein, Western blot analysis was performed (Figure 5.14) but as we could see there are significant level of protein expression. Taken together, these data shows that carboxyl group at position 149 is essential to the ammonium transport activity of Amt-1 but full conclusion will need to combine structural studies which still have to be performed for these mutants.

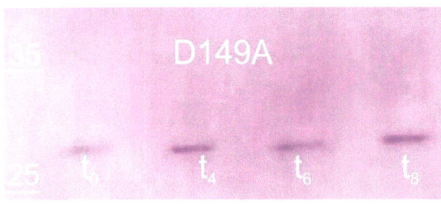

Figure 5.14: Amt-1 variant expression in *ΔamtB* C43(DE3) *E.coli* strain using tetra-His antibody against C-terminus hexa-His-tag. M-protein marker; t_0-, t_4-, t_6-, t_8- zero, four, six and eight hours after adding 1 mM of NH_4Cl into the culture medium, respectively.

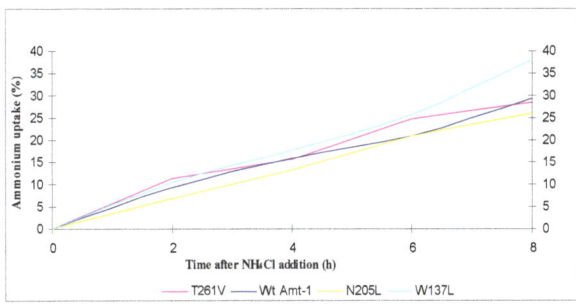 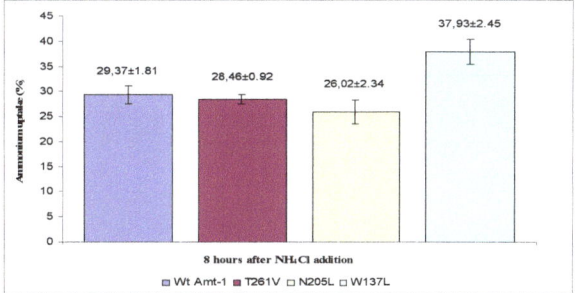

Figure 5.15: *In vivo* **determination of ammonium uptake in** *ΔamtB* **C43(DE3)** *E. coli* **strains.** Comparison of the *in vivo* ammonium uptake capacity between plasmid carrying the wild-type *A. fulgidus amt-1* gene and plasmids carrying different variants of *amt-1* (T261V, N205L, W137L, respectively) following transformation in BL(21)C43 *E.coli* cells that were knocked-out in their *amtB* gene.The graphic on the left shown ammonium uptake within 8 hours after addition of 1 mM NH_4Cl into minimal medium. The graphic on the right s shown in the histogram representation of ammonium uptake at a chosen point time (8 hours) after NH_4Cl addition. Values are means ± S.D (n=12).

We have also studied the substitution of the highly conserved tryptophane 137 residue for a leucin at the possible substrate recruitment and selectivity side (Figure 5.15). A strain expressing W137L accumulates larger amount of ammonium (37.93 % ± 2.45) than a plasmid carrying the *A. fulgidus amt*-1 (29.37 % ± 1.81).

Western blot analysis (Figure 5.16) indicated that the W137L protein was present in the same or even lower amounts as a plasmid carrying the *A. fulgidus amt*-1. Protein amounts were not elevated and hence higher activity was not simply due to increased protein expression. Substitution of W137 to H (29.56 % ± 0.59) did not reflect on transport activity in comparison to a plasmid carrying the *A. fulgidus amt*-1 (31.5 % ± 0.82) (Figure 5.11).

Taken together, our results indicate that the highly conserved W137 residue may not be required for recruitment of the $NH4^+$ ion.

Although structural data showed some conformational changes in surrounding F96 and F204 the ammonium uptake assay showed not significant changes in uptake capacities (28.46 % ± 0.92 for T261V and 26.02 % ± 2.34 for N205L) in comparison to a plasmid carrying the *A. fulgidus amt*-1 (29.37 % ± 1.81).

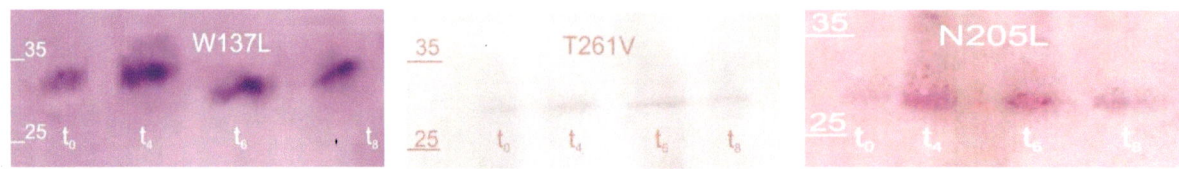

Figure 5.16: Amt-1 variant expression in _ΔamtB_ C43 (DE3) _E. coli_ strains using tetra-His antibody against C-terminal hexa-His-tag. M-protein marker; t_0-, t_4-, t_6-, t_8- zero, four, six and eight hours after adding 1 mM of NH_4Cl into the culture medium, respectively.

6. References

1. **Khademi S., O'Connell J., Remis J., Robles-Colmenares Y., Miercke L. J., Stroud R. M**. (2004). Mechanism of ammonia transport by Amt/MEP/Rh: Structure of AmtB at 1.5 Å. *Science*. 305: 1587-1594.

2. **Zheng L., Kostrewa D., Berneche S., Winkler F. K., and Li X. D**. (2004). The mechanism of ammonia transport based on the crystal structure of AmtB of *E. coli*. *PNAS*. 101: 17090-17095.

3. **Andrade S. L. A., Dickmanns A., Ficner R., and Einsle O**. (2005). Crystal structure of the archaeal ammonium transporter Amt-1 from *Archaeoglobus fulgidus*. *PNAS*. 102 (42): 14994-9.

4. **Domenico Lupo, Xiao-Dan Li, Anne Durand, Takashi Tomizaki, Baya Cherif-Zahar, Giorgio Matassi, Mike Merrick, and Fritz K. Winkler.** The 1.3Å resolution structure of *Nitrosomonas europaea* Rh50 and mechanistic implications for NH_3 transport by Rhesus family proteins. *PNAS*. 104 (49): 19303–19308.

5. **Li X., Jayachandran S., Nguyen H. H., Chan M. K**. (2007). Crystal structure of the *Nitrosomonas europaea* Rh protein. *PNAS*. 104: 19279-19284.

6. **Franz Gruswitz, Sarika Chaudhary, Joseph D. Ho, Avner Schlessinger, Bobak Pezeshki, Chi-Min Ho, Andrej Sali, Connie M. Westhoff, and Robert M. Stroud.** (2010). Function of human Rh based on structure of RhCG at 2.1 Å. *PNAS*. 21: 9638.

7. **Rees D. C. and Howard J. B**. (2000). Nitrogenase: standing at the crossroads. *Curr. Opin. Chem. Biol*. 4: 559-566.

8. **Susan M. Howitt and Michael K. Udvardi.** (2000). Structure, function and regulation of ammonium transporters in plants. *Biochimica et Biophysica Acta - Biomembranes*. 1465 (1-2): 152-170.

9. **Kleiner D.** (1981). The transport of NH_3 and NH_4^+ across biological membranes. *Biochim. Biophys. Act.* 639: 41-52.

10. **Susan W. Gay and Katharine F. Knowlton.** (2009). Ammonia Emissions and Animal Agriculture. *Virginia Cooperative Extension.* 442: 110-116.

11. **Hackette S. L., Skye G. E., Burton C, Segel I. H.** (1970). Characterization of an ammonium. Transport system in filamentous fungi with methyl ammonium as the substrate. *JBC.* 245: 4241-50.

12. **Ninnemann O., Jauniaux J. C., and Frommer W. B.** (1994). Identification of a high affinity NH_4^+ transporter from plants. *EMBO J.* 13: 3464-3471.

13. **Marini A. M., Vissers S., Urrestarazu A., and Andre B.** (1994). Cloning and expression of the MEP1 gene encoding an ammonium transporter in *Saccharomyces cerevisiae. EMBO J.* 13: 3456–3463.

14. **Marini A. M., Soussi-Boudekou S., Vissers S., and Andre B.** (1997). A family of ammonium transporters in *Saccharomyces cerevisiae. Mol. Cell. Biol.* 17 (8): 4282-4293.

15. **Blakey D., Leech A., Thomas G. H., Coutts G., Findlay K., Merrick M.** (2002). Purification of the *E. coli* ammonium transporter AmtB reveals a trimeric stoichiometry. *Biochem. J.* 364: 527-535.

16. **Andrade S. L. A. and Einsle O.** (2007). The Amt/Mep/Rh family of ammonium transport proteins (Review). *Molecular membrane biology.* 24 (5-6): 357–365.

17. **Marini A. M., Urrestarazu A., Beauwens R., Andre B.** (1997). The Rh (Rhesus) blood group polypeptides are related to NH_4^+ transporters. *Trends Biochem Sci.* 22: 460–461.

18. **Marini A. M., Matassi G., Raynal V., Andre B., Cartron J. P. and Cherif-Zahar B.** (2000). The human Rhesus-associated RhAG protein and a kidney homologue promote ammonium transport in yeast. *Nat. Genet.* 26: 341–344.

19. **Huang C. H. and Liu P. Z.** (2001). New insights into the Rh superfamily of genes and proteins in erythroid cells and nonerythroid tissues. *Blood Cells Mol. Dis.* 27: 90–101.

20. **Liu Z., Peng J., Mo R., Hui C., and Huang C. H.** (2001). Rh Type B Glycoprotein is a n new member of the Rh Superfamily and a putative ammonia transporter in mammals.*J. Biol. Chem.* 276: 1424–1433.

21. **Liu Z., Chen Y., Mo R., Hui C., Cheng J. F., Mohandas N. and Huang, C. H.** (2000). Characterization of Human RhCG and Mouse Rhcg as Novel Nonerythroid Rh Glycoprotein Homologues Predominantly Expressed in Kidney and Testis. *J. Biol. Chem.* 275: 25641–25651.

22. **Westhoff C. M., Ferreri-Jacobia M., Mak D. O., and Foskett J. K.** (2002). Identification of the erythrocyte Rh blood group glycoprotein as a mammalian ammonium transporter. *J Biol Chem.* 277: 12499-12502.

23. **Ludewig U.** (2006). Ion transport versus gas conduction: function of AMT/Rh-type proteins. *Transfus Clin Biol.* 13 (1-2):111-6.

24. **Ludewig U.** (2004). Electroneutral ammonium transport by basolateral Rhesus B glycoprotein. *J Physiol.* 559 (3): 751-9.

25. **Mayer M., Schaaf G., Mouro I., Lopez C., Colin Y., Neumann P., Cartron J. P., Ludewig U.** (2006). Different transport mechanisms in plant and human AMT/Rh-type ammonium transporters. *J Gen Physiol.* 127 (2):133-44.

26. **Soupene E., Inwood W. and Kustu S.** (2004). Lack of the Rhesus protein Rh1 impairs growth of the green alga *Chlamydomonas reinhardtii* at high CO_2 *PNAS.* 101: 7787–7792.

27. **Matassi G., Cherif-Zahar B., Pesole G., Raynal V., and Cartron J. P.** (1999). The members of the RH gene family (RH50 and RH30) followed different evolutionary pathways. *J. Mo.l Evol.* 48: 151-159.

28. **Kitano T., and Saitou N.** (2000). Evolutionary history of the Rh blood group-related genes in vertebrates. *Immunogenetics.* 51: 856-862.

29. **Ludewig U., von Wirén N., Rentsch D., and Frommer W. B.** (2001). *Genome Biology.* 2 (3):1010.1-1010.5

30. **La Paglia Christopher and Patricia L. Hartzell.** (1997). Stress-Induced Production of Biofilm in the hyperthermophile *Archaeoglobus fulgidus.* *Applied and Environmental Microbiology* 63 (8): 3158-3163.

31. **Klenk H. P., Clayton R. A., Tomb J., White O., Nelson K. E., Ketchum K. A., Dodson R. J., Gwinn M., Hickey E. K., Peterson J. D., Richardson D. L., Kerlavage A. R., Graham D. E., Kyrpides N. C., Fleischmann R. D., Quackenbush J., Lee N. H., Sutton G. G., McKenney K., Adams M. D., Loftus B., Peterson S., Reich C. I., McNeil L. K., Badger J. H., Glodek A., Zhou L., Overbeek R., Gocayne J. D., Weidman J. F., McDonald L., Utterback T., Cotton M. D., Spriggs T., Artiach P., Kaine B. P., Sykes M. S., Sadow P. W., Andrea K. P., Bowman C., Fujii C., Garland S. A., Mason T. M., Olsen G. J., Fraser C. M., Smith H. O., Woese C. R. & Venter C.** (1997). The complete genome sequence of the hyperthermophilic, sulphate-reducing archaeon *Archaeoglobus fulgidus. Nature.* 390: 364- 370.

32. **Woese C. R., Achenbach L., Rouvière P. and Mandelco L.** (1991). Archael phylogeny: reexaminiation of the phylogenetic position of *Archaeoglobus fulgidus* in light of certain composition-induced artifacts. *Syst. Appl. Microbiol.* 14: 364-371.

33. **von Wirén N., Lauter F. R., Ninnemann O., Gillissen B., Walch-Liu P., Engels C., Jost W., Frommer W. B.** (2000). Differential regulation of three functional

ammonium transporter genes by nitrogen in root hairs and by light in leaves of tomato. *Plant J.* 21(2): 167-75.

34. **Galagan J. E., Nusbaum C., Roy A., Endrizzi M. G., MacDonald P., FritzHugh W., Calvo S., Engels R., Smirnov S., Atmoor D., Brown A., Allen N., Naylor J., Stange-Thormann N., DeArcellano K., Johnson R., Linton L., McEwan P., McKernan K., Talamas J., Tirrell A., Ye W., Zimmer A., Barber R. D., Cann I., Graham D. E., Graham D. A., Guss A. M., Hedderich R., Ingram-Smith C., Kuettner H. C., Krzycki J. A., Leigh J. A., Li W., Liu J., Mukhopadhyay B., Reeve J. N., Smith K., Springer T. A., Umayam L. A., White O., White R. H., Conway de Macario E., Ferry J. G., Jarrell K. F., Jing H., Macario A. J., Paulsen I., Pritchett M., Sowers K. R., Swanson R.V., Zinder S. H., Lander E., Metcalf W. W. & Birren B.** (2002). The genome of *M. acetivorans* reveals extensive metabolic and physiological diversity. *Genome Res.* 12: 532-542.

35. **Montesinos M. L., Muro-Pastor A. M., Herrero A. and Flores E.** (1998). Ammonium/methylammonium permeases of *Cyanobacterium*. Identification and analysis of the three nitrogen-regulated amt genes in synechocystis sp. PCC 6803. *J. Biol. Chem.* 273: 31461-31470.

36. **Coutts G., Thomas G., Blakey D., Merrick M.** (2002). Membrane sequestration of the signal transaduction protein GlnK by the ammonium transporter AmtB. *EMBO J.* 21: 536-545.

37. **Loque D., Lalonde S., Looger L., Wiren N., and Frommer W. B. (2007).** A cytosolic transactivation domain is essential for ammonium uptake. *Nature.* 446: 195-198.

38. **Heijne G. and Gavel Y.** (1988). Topogenic signals in integral membrane proteins. *FEBS Journal.* 174: 671-678.

39. **Abramson J., Smirnova I., Kasho V., Verner G., Kaback H. R. and Iwata S.** (2003). Structure and mechanism of the lactose permease of *Escherichia coli. Science.* 301: 610- 615.

40. **Dutzler R., Campbell E. B., Cadene M., Chait B. T. and MacKinnon R.** (2002). X-ray structure of a ClC chloride channel at 3.0 Å reveals the molecular basis of the anion selectivity. *Nature.* 415: 287-294.

41. **Kleiner D.** (1985). Bacterial ammonium transport. *FEMS Microbiol. Rev.* 32: 87-100.

42. **Gazzarrini S., Lejay L., Gojon A., Ninnemann O., Frommer W. B., and von Wiren N.** (1999). Three functional transporters for constitutive, diurnally regulated, and starvation-induced uptake of ammonium into *Arabidopsis* roots. *Plant Cell.* 11: 937-948.

43. **Ludewig U., Wiren N. and Frommer W. B.** (2002). Uniport of NH4$^+$ by the root hair plasma membrane ammonium transporter LeAMT1;1. *J. Biol. Chem.* 277: 13548-13555.

44. **Nakhoul N. L., Dejong H., Abdulnour-Nakhoul S. M., Boulpaep E. L., Hering-Smith K., and Hamm L. L.** (2005). Characteristics of renal Rhbg as an NH4$^+$ transporter. *Am. J. Physiol. Ren. Physiol.* 288: F170-F181.

45. **Soupene E., He L., Yan D. and Kustu S.** (1998). Ammonia acquisition in enteric bacteria: Physiological role of the ammonium/methylammonium transport B (AmtB) protein. *PNAS.* 95: 7030–7034.

46. **Soupene E., Lee H. and Kustu S.** (2002). Ammonium/methylammonium transport (Amt) proteins facilitate diffusion of NH$_3$ bidirectionally. *PNAS.* 99: 3926–3931.

47. **Arnaud Javelle, Domenico Lupo, Xiao-Dan Li, Mike Merrick, Mohamed Chami, Pierre Ripoche, Fritz K. Winkler.** (2007). Structural and mechanistic aspects of Amt/Rh proteins. *Journal of Structural Biology.* 158: 472–481.

48. **Dev T. Britto and Herbert J. Kronzuker.** (2002). NH4$^+$ toxicity in higher plants: a critical review. *J. Plant Physiol.* 159: 567 – 584.

49. **Jeremy M. Berg, John L. Tymoczko, and Lubert Stryer.** (2002). *Biochemistry.* 5th ed. New York: W. H. Freeman.

50. **Arcondeguy T, Jack R., Merrick M.** (2001). P_{II} signal transduction proteins, pivotal players in microbial nitrogen control. *Mol. Biol. Rev.* 65: 80-108.

51. **Jack R., de Zamaroczy M., and Merrick M.** (1999). The Signal Transduction Protein GlnK is required for NifL-Dependent Nitrogen Control of nif Gene Expression in *Klebsiella pneumoniae. J. Bacteriol.* 181: 1156-1162.

52. **Little R., Colombo V., Leech A., Dixon R.** (2002). Direct interaction of the NifL regulatory protein with the GlnK signal transducer enables the *Azotobacter vinelandii* NifL-NifA regulatory system to respond to conditions replete for nitrogen. *J Biol Chem.* 277 (18): 15472-81.

53. **Zhang Y., Pohlmann E. L., Ludden P. W., Roberts G. P.** (2001). Functional characterization of three GlnB homologs in the photosynthetic bacterium *Rhodospirillum rubrum*: roles in sensing ammonium and energy status. *J Bacteriol.* 183 (21): 6159-6168.

54. **Gruswitz F., O'Connell J., Stroud R. M.** (2007). Inhibitory complex of the transmembrane ammonia channel, AmtB, and the cytosolic regulatory protein GlnK at 1.96 Å. *PNAS.* 104: 42-47.

55. **Conroy M. J., Durand A., Lupo S., Li X. D., Bullough P. A., Winkler F. K and Merrick M.** (2007). The crystal structure of the *E.Coli* AmtB-GlnK complex revels how GlnK regulates the ammonia channel. *PNAS.* 104: 1213-1218.

56. **Yildiz O., Kalthoff C., Raunser S., Kühlbrandt W.** (2007). Structure of GlnK1 with bound effectors indicates regulatory mechanism for ammonia uptake. *EMBO J.* 26: 589-599.

57. **Miroux B., Walker J. E.** (1996). Over-production of proteins *in Escherichia coli*: mutant hosts that allow synthesis of some membrane proteins and globular proteins at high levels. *J Mol Biol.* 260 (3): 289-298.

58. **TargeTron™ Gene Knockout System.** User guide-Sigma Aldrich. Catalog Number TA0100.

59. **Perutka J.** (2004). Use of computer-designed group II introns to disrupt *Escherichia coli* DExH/D-box protein and DNA helicase genes. *J. Mol. Biol.* 336: 421–439.

60. **Ingle D. J., and Crouch S. R.** (1988). Book: Spectrochemical Analysis, Prentice Hall, New Jersey.

61. **Sambrook and Russell.** (2001). *Molecular Cloning: A Laboratory Manual* (3rd ed.). Validity of nucleic acid purities monitored by 260nm/280nm absorbance ratios. *BioTechniques* 18: 62–63.

62. **Quick-change Site-directed Mutagenesis Kit.** Instruction manual. Catalog Nr: 200518. Strategene.

63. **Wilson K. and Walker J.** (2005). Book: Principles and Techniques of Practical Biochemistry, *Cambridge University Press*, 5th edition.

64. **Rehm H.** (2006). Der Experimentator, Proteinbiochemie/Proteomics. *Spektrum Akademischer Verlag.* 5 Auflage.

65. **Instructions for Micro BCA Protein Assay Kit.** Thermo Scientific. Prod.Nr: 23235

66. **Thomas Marshall and Katherine M. Williams.** (2004). Drug interference in the Bradford and 2, 2'-bicinchoninic acid protein assays. *Analytical Biochemistry.* 198 (2): 352-354.

67. **Towbin H., Staehelin T., Gordon J.** (1979). Electrophoretic transfer of proteins from polyacrylamide gels to nitrocellulose sheets: procedure and some applications. *PNAS*. 76 (9): 4350–4354

68. **Varner J. E., Bulen W. A., Vanecko Steve, R., Burrell C.** (1953). Determination of Ammonium, Amide, Nitrite, and Nitrate Nitrogen in Plant Extracts. *Anal. Chem.* 25 (10): 1528–1529.

69. **Shinji Wakisaka, Takashi Tachiki, Ha-Chin Sung, Hidehiko Kumagai, Tatsurokuro Tochikura, and Susumu Matsui.** (1987). A rapid assay method for ammonia using glutamine synthetase from glutamate-producing bacteria. *Analytical Biochemistry*. 163 (1): 117-122.

70. **Khramov V. A.** (1982). Determination of the carbamate kinase activity of bacteria. *Mikrobiologiia*. 51 (6): 1002-5.

71. **Cohen N. S., Kyan F. S., Cheung, C. W., and Raijman L..** (1985). The apparent Km of ammonia for carbamoyl phosphate synthetase (ammonia) in situ. *Biochem. J.* 229 (1): 205–211.

72. **Reid H. Leonard.** (1961). Quantitative Range of Nessler's Reaction with Ammonia. *Clinical Chemistry*. 9: 417-422.

73. **T. Salzer.** *Fresenius' journal of analytical chemistry*. 20 (1): 225-231

74. **Ferris H. Venette R. C., H. R. van der Meulen and S. S. Lau.** (1998). Nitrogen mineralization by bacterial-feeding nematodes: verification and measurement. *Plant and Soil*. 203: 159–171.

75. **Hahn Theo** (2002). International Tables for Crystallography, Volume A: Space Group Symmetry, A (5th ed.), Berlin, New York: Springer-Verlag.

76. **Neer Asherie.** (2004). Protein crystallization and phase diagrams. *Methods* 34: 266–272.

77. **Naomi E Chayen and Emmanuel Saridakis.** (2008). Protein crystallization: from purified protein to diffraction-quality crystal. *Nature Methods.* 152 (5):147-153.

78. **Bragg W. H.** (1908). The nature of γ- and X-rays. *Nature.* 77: 270.

79. **Leslie A.** (1993). Data collection and processing. Proceedings of the CCP4 study weekennd. Warrington, UK: SERC Daresbury Laboratory. 44–51.

80. **Kabsch W.** (1998). Automatic indexing of rotation diffraction patterns. *Journal of Applied Crystallography.* 21: 67–71.

81. **Leslie A.** (1999). Integration of macromolecular diffraction data. *Acta Cryst.* D55: 1696-1702.

82. **Miroslav Z. Papiz and Graeme Winter** (2010). Biomolecular X-Ray Crystallography, Structure Determination Methods. *Encyclopedia of Spectroscopy and Spectrometry (Second Edition)*: 185-193.

83. **Gale Rhodes.** (2006). Book. Crystallography Made Crystal Clear. A Guide for Users of Macromolecular Models. Elsevier LTD, Oxford.

84. **Jan Drenth**. (1999). Book. Principles of Protein X-ray Crystallography.Second edition. Springer-Verlag New York.

85. **Bracewell R. N.** (2000). Book. The Fourier Transform and Its Applications. The 3rd edition. Boston: McGraw-Hill.

86. **Ten Eyck L. F.** (1973). Crystallographic fast Fourier transforms. *Acta Crystallogr*. A. A29: 183–91.

87. **The CCP4 suite: programs for protein crystallography.** (1994). *Acta Crystallogr D Biol Crystallogr.* 1 (50):760-763.

88. **Terwilliger T. C.** (1994). MAD Phasing: Treatment of Dispersive Differences as Isomorphous Replacement Information. *Acta Cryst.* D50: 17–23.

89. **Vagin, A. A., Steiner, R. S., Lebedev, A. A., Potterton L., McNicholas S., Long F., and Murshudov, G. N.** (2004). REFMAC5 dictionary: organisation of prior chemical knowledge and guidelines for its use. *Acta Cryst.* D60: 2284-2295.

90. **Brunger A. T.** (1992). Free R value: a novel statistical quantity for assessing the accuracy of crystal structures. *Nature.* 355 (6359): 472–475.

91. **Arnaud Javelle, Domenico Lupo, Lei Zheng, Xiao-Dan Li, Fritz K. Winkler and Mike Merrick.** (2006). An Unusual Twin-His Arrangement in the Pore of Ammonia Channels Is Essential for Substrate Conductance *JBC.* 281: 39492-39498.

92. **Anna Maria Marini, Merlanie Boeckstaens, Fatine Benjelloun, Baya Cherrif-Zahar and Bruno Andre.** (2006). Structural involvement in substrate recognition of an essential aspartate residue conserved in Mep/Amt and Rh-type ammonium transporters. *Curr Genet.* 49: 364–374.

93. **Thompson J. D, Gibson T. J, Higgins D. G.** (2002). Multiple sequence alignment using ClustalW and ClustalX. *Curr Protoc Bioinformatics.* Chapter 2: Unit 2.3.

94. **Luis Caspeta, Noemí Flores, Néstor O. Pérez, Francisco Bolívar, Octavio T. Ramírez.** (2009). The effect of heating rate on *Escherichia coli* metabolism, physiological stress, transcriptional response, and production of temperature-induced recombinant protein: A scale-down study. *Biotechnology and Bioengineering.* 102 (2): 468-482.

95. **Hansen L. H, Knudsen S, Sørensen S. J.** (1998). The effect of the *lacY* gene on the induction of IPTG inducible promoters, studied in *Escherichia coli* and *Pseudomonas fluorescens*. *Curr. Microbiol.* 36 (6): 341-347.

96. **Andrade S. L. A., Dickmanns A., Ficner R., and Einsle O.** (2005). Expression, purification and crystallization of the ammonium transporter Amt-1 from *Archaeoglobus fulgidus. Acta Cryst.* F61: 861-863.

97. **Gilbert G. Privé.** (2007). Detergents for the stabilization and crystallization of membrane proteins. *Structural Biology of Membrane Proteins.* 41 (4): 388-397.

98. **Brungerm A. T.** (1993). Assessment of phase accuracy by cross validation: the free R value. Methods and applications. *Acta Crystallogr.* D49: 24-36.

99. **Emsley P and Cowtan K.** (2004). Coot: model-buliding tools for molecular graphics. *Acta Crystallograph.* D60: 2126-2132.

100. **Anscombe, Francis J.** (1973). Graphs in statistical analysis. *The American Statistician* 27: 17–21.

101. **Copper V. S., Bennet A. F., Lenski R. E.** (2001). Evolution of thermal dependence of growth rate of *Escherichia coli* population during 20,000 generations in a constant environment. *Evolution.*(5) 55: 889- 896.

7. Appendix

7.1. DNA and amino acids sequence of the Amt-1

```
 M   S   D   G   M   V   A   W   I   L   A   S   T   A   L   V   M   L   M   V
ATG AGT GAC GGA AAT GTC GCA TGG ATA CTC GCA TCC ACG GCC CTT GTA ATG CTG ATG GTG

 P   G   V   G   F   F   Y   A   G   M   V   R   R   K   N   A   V   N   M   I
CCG GGA GTG GGG TTC TTT TAC GCA GGA ATG GTA AGG AGA AAG AAT GCA GTT AAC ATG ATT

 A   L   S   F   I   S   L   I   I   T   V   L   L   W   I   F   Y   G   Y   S
GCG CTG AGC TTC ATA TCA CTC ATA ATC ACG GTT TTG CTG TGG ATA TTC TAC GGC TAC TCG

 V   S   F   G   N   D   I   S   G   I   I   G   G   L   M   Y   A   L   L   S
GTG AGC TTC GGA AAT GAC ATC TCT GGA ATC ATT GGA GGG CTG AAT TAT GCA CTG CTA AGC

 G   V   K   G   G   E   D   L   L   F   M   M   Y   Q   M   M   F   A   A   V
GGA GTT AAG GGG GAG GAT TTG CTG TTC ATG ATG TAC CAG ATG ATG TTC GCC GCT GTC ACA

 T   I   A   I   L   T   S   A   I   A   E   R   A   K   V   S   S   F   I   L
ATT GCA ATC CTC ACC TCC GCA ATT GCT GAG AGA GCA AAA GTT TCA TCG TTC ATT CTC CTC

 S   A   L   W   L   T   F   V   Y   A   P   F   A   H   W   L   W   G   G   G
AGC GCT CTG TGG CTT ACG TTC GTT TAC GCC CCC TTC GCA CAC TGG CTT TGG GGT GGG GGG

 W   L   A   K   L   G   A   L   D   F   A   G   G   M   V   V   H   I   S   S
TGG CTG GCA AAG CTC GGC GCC CTC GAC TTT GCT GGA GGT ATG GTT GTT CAC ATA AGC TCG

 G   F   A   A   L   A   V   A   M   T   I   G   K   R   A   G   F   E   E   Y
GGA TTT GCT GCA CTT GCA GTC GCG ATG ACG ATA GGT AAG AGG GCG GGA TTC GAG GAG TAC

 S   I   E   P   H   S   I   P   L   T   L   I   G   A   A   L   L   W   F   G
TCG ATA GAG CCA CAC AGC ATT CCG CTG ACG CTC ATT GGC GCT GCC CTG CTT TGG TTT GGG

 W   F   G   F   N   G   G   S   A   L   A   A   N   D   V   A   I   N   A   V
TGG TTC GGA TTC AAC GGC GGA AGT GCA TTG GCT GCA AAC GAT GTG GCC ATC AAC GCC GTG

 V   V   T   N   T   S   A   A   V   A   G   F   V   W   M   V   I   G   W   I
GTG GTC ACA AAC ACC TCA GCA GCA GTA GCA GGG TTT GTC TGG ATG GTA ATT GGA TGG ATT

 K   G   K   P   G   S   L   G   I   V   S   G   A   I   A   G   L   A   A   I
AAG GGA AAG CCG GGG AGT CTT GGG ATA GTG AGC GGT GCA ATT GCT GGG CTT GCC GCC ATA

 T   P   A   A   G   F   V   D   V   K   G   A   I   V   I   G   L   V   A   G
ACC CCC GCA GCA GGC TTT GTG GAT GTA AAG GGA GCG ATT GTC ATA GGT CTT GTG GCT GGA

 I   V   C   Y   L   A   M   D   F   R   I   K   K   K   I   D   E   S   L   D
ATA GTA TGC TAC CTT GCT ATG GAC TTC AGA ATA AAG AAG AAG ATA GAC GAG AGC CTT GAT

 A   W   A   I   H   G   I   G   G   L   W   G   S   V   A   V   G   I   L   A
GCT TGG GCG ATT CAC GGA ATA GGC GGT TTA TGG GGA AGT GTT GCA GTT GGC ATT CTT GCA

 N   P   E   V   N   G   Y   A   G   L   L   F   G   N   P   Q   L   L   V   S
AAT CCG GAG GTT AAC GGA TAT GCA GGC CTA CTG TTC GGA AAT CCG CAA CTG CTA GTT TCA

 Q   L   I   A   V   A   S   T   T   A   Y   A   F   L   V   T   L   I   L   A
CAA CTG ATT GCG GTT GCA TCC ACA ACA GCC TAC GCC TTC CTC GTG ACG CTG ATA CTG GCA

 K   A   V   D   A   A   V   G   L   R   V   S   S   Q   E   E   Y   V   G   L
AAG GCT GTT GAT GCC GCT GTG GGG CTG AGG GTT AGC TCG CAG GAG GAG TAC GTC GGT CTC

 D   L   S   Q   H   E   E   V   A   Y   T   stop
GAC CTG TCG CAG CAT GAG GAG GTT GCC TAC ACG TGA
```

7.2. Amino acids, their single-letter data-base codes (SLC), and their corresponding DNA codons

Amino Acid	SLC	DNA codons
Isoleucine	I	ATT, ATC, ATA
Leucine	L	CTT, CTC, CTA, CTG, TTA, TTG
Valine	V	GTT, GTC, GTA, GTG
Phenylalanine	F	TTT, TTC
Methionine	M	ATG
Cysteine	C	TGT, TGC
Alanine	A	GCT, GCC, GCA, GCG
Glycine	G	GGT, GGC, GGA, GGG
Proline	P	CCT, CCC, CCA, CCG
Threonine	T	ACT, ACC, ACA, ACG
Serine	S	TCT, TCC, TCA, TCG, AGT, AGC
Tyrosine	Y	TAT, TAC
Tryptophan	W	TGG
Glutamine	Q	CAA, CAG
Asparagine	N	AAT, AAC
Histidine	H	CAT, CAC
Glutamic acid	E	GAA, GAG
Aspartic acid	D	GAT, GAC
Lysine	K	AAA, AAG
Arginine	R	CGT, CGC, CGA, CGG, AGA, AGG
Stop codons	Stop	TAA, TAG, TGA

7.3. DNA sequence alignment of wild type *amt-1* from *Archaeoglobus fulgidus* with designed mutants of *amt-1* from *Archaeoglobus fulgidus*.

Designed mutations are highlighted in red and the hexa-histidine affinity tag on C-terminus is highlighted in yellow. Fwd-Forward Primer, Rev-Reverse Primer

A258T

Scores table

```
SeqA Name        Len(nt)  SeqB Name        Len(nt)  Score
=========================================================
1    wt-amt1     1176     2    A258T-fwd    1054     94
1    wt-amt1     1176     3    A258T-rev    1104     89
2    A258T-fwd   1054     3    A258T-rev    1104     77
=========================================================
```

Alignment

```
wt-amt1     --------------------------------------------------------ATGAGTGAC 9
A258T-fwd   GTGAGAATTCCCTCTGAATATTTTGTTTAACTTTAAGAAGGAGATATACATATGAGTGAC 60
A258T-rev   ------------------------------------------------------------

wt-amt1     GGAAATGTCGCATGGATACTCGCATCCACGGCCCTTGTAATGCTGATGGTGCCGGGAGTG 69
A258T-fwd   GGAAATGTCGCATGGATACTCGCATCCACGGCCCTTGTAATGCTGATGGTGCCGGGAGTG 120
A258T-rev   ------------------------------------------------------------

wt-amt1     GGGTTCTTTTACGCAGGAATGGTAAGGAGAAAGAATGCAGTTAACATGATTGCGCTGAGC 129
A258T-fwd   GGGTTCTTTTACGCAGGAATGGTAAGGAGAAAGAATGCAGTTAACATGATTGCGCTGAGC 180
A258T-rev   ------------------------------------------------------------

wt-amt1     TTCATATCACTCATAATCACGGTTTTGCTGTGGATATTCTACGGCTACTCGGTGAGCTTC 189
A258T-fwd   TTCATATCACTCATAATCACGGTTTTGCTGTGGATATTCTACGGCTACTCGGTGAGCTTC 240
A258T-rev   ----------------TGGGGTATTTCTTACGGGCTAACTCCGGT-----GGAGCTTTCG 39

wt-amt1     GGAAAT-GACATCTC--TGGAATCA-TTGGAGGG--CTGAATTA--TGCACTGCTAA-GC 240
A258T-fwd   GGAAAT-GACATCTC--TGGAATCA-TTGGAGGG--CTGAATTA--TGCACTGCTAA-GC 291
A258T-rev   GGAAATTGACATTCTCTTGGAATCAATTGGAAGGGATTGAATTTATTGAACTGCTAAAGC 99

wt-amt1     GGAG-TTAAGGGGGAGGA--TTTGCTGTTCA-TGATGTACC-AGATGATGTTCG--CCGC 293
A258T-fwd   GGAG-TTAAGGGGGAGGA--TTTGCTGTTCA-TGATGTACC-AGATGATGTTCG--CCGC 344
A258T-rev   GGAGGTTAAGGGGGAAGGATTTTGTTGTTCAATGATGTACCCAGATGATGTTTCGCCCGC 159

wt-amt1     TGTCACAATTGC-AATCCTCACCTCC-GCAATTGCTGAGAGAGCAAAA-GTTTCATCGTT 350
A258T-fwd   TGTCACAATTGC-AATCCTCACCTCC-GCAATTGCTGAGAGAGCAAAA-GTTTCATCGTT 401
A258T-rev   TGTCACAATTTCTGATTCTCACCTCCCGCAATTGCTGAGAGAGCAAAAAGTTTCATCGTT 219

wt-amt1     CATTCTCCTCAGCGCTCTGTGGCTTACGTTCGTTTACGCCCCCTTCGCACACTGGCTTTG 410
A258T-fwd   CATTCTCCTCAGCGCTCTGTGGCTTACGTTCGTTTACGCCCCCTTCGCACACTGGCTTTG 461
A258T-rev   CATTCTCCTCAGCGCTCTGTGGCTTACGTTCGTTTACGCCCCCTTCGCACACTGGCTTTG 279

wt-amt1     GGGTGGGGGGTGGCTGGCAAAGCTCGGCGCCCTCGACTTTGCTGGAGGTATGGTTGTTCA 470
A258T-fwd   GGGTGGGGGGTGGCTGGCAAAGCTCGGCGCCCTCGACTTTGCTGGAGGTATGGTTGTTCA 521
A258T-rev   GGGTGGGGGGTGGCTGGCAAAGCTCGGCGCCCTCGACTTTGCTGGAGGTATGGTTGTTCA 339

wt-amt1     CATAAGCTCGGGATTTGCTGCACTTGCAGTCGCGATGACGATAGGTAAGAGGGCGGGATT 530
A258T-fwd   CATAAGCTCGGGATTTGCTGCACTTGCAGTCGCGATGACGATAGGTAAGAGGGCGGGATT 581
A258T-rev   CATAAGCTCGGGATTTGCTGCACTTGCAGTCGCGATGACGATAGGTAAGAGGGCGGGATT 399

wt-amt1     CGAGGAGTACTCGATAGAGCCACACAGCATTCCGCTGACGCTCATTGGCGCTGCCCTGCT 590
```

```
A258T-fwd   CGAGGAGTACTCGATAGAGCCACACAGCATTCCGCTGACGCTCATTGGCGCTGCCCTGCT 641
A258T-rev   CGAGGAGTACTCGATAGAGCCACACAGCATTCCGCTGACGCTCATTGGCGCTGCCCTGCT 459


wt-amt1     TTGGTTTGGGTGGTTCGGATTCAACGGCGGAAGTGCATTGGCTGCAAACGATGTGGCCAT 650
A258T-fwd   TTGGTTTGGGTGGTTCGGATTCAACGGCGGAAGTGCATTGGCTGCAAACGATGTGGCCAT 701
A258T-rev   TTGGTTTGGGTGGTTCGGATTCAACGGCGGAAGTGCATTGGCTGCAAACGATGTGGCCAT 519


wt-amt1     CAACGCCGTGGTGGTCACAAACACCTCAGCAGCAGTAGCAGGGTTTGTCTGGATGGTAAT 710
A258T-fwd   CAACGCCGTGGTGGTCACAAACACCTCAGCAGCAGTAGCAGGGTTTGTCTGGATGGTAAT 761
A258T-rev   CAACGCCGTGGTGGTCACAAACACCTCAGCAGCAGTAGCAGGGTTTGTCTGGATGGTAAT 579


wt-amt1     TGGATGGATTAAGGGAAAGCCGGGGAGTCTTGGGATAGTGAGCGGTGCAATTGCTGGGCT 770
A258T-fwd   TGGATGGATTAAGGGAAAGCCGGGGAGTCTTGGGATAGTGAGCGGTGCAATTGCTGGGCT 821
A258T-rev   TGGATGGATTAAGGGAAAGCCGGGGAGTCTTGGGATAGTGAGCGGTGCAATTGCTGGGCT 639


wt-amt1     TGCCGCCATAACCCCCGCAGCAGGCTTTGTGGATGTAAAGGGAGCGATTGTCATAGGTCT 830
A258T-fwd   T<mark style="background:orange">ACC</mark>GCCATAACCCCCGCAGCAGGCTTTGTGGATGTAAAGGGAGCGATTGTCATAGGTCT 881
A258T-rev   TGCCGCCATAACCCCCGCAGCAGGCTTTGTGGATGTAAAGGGAGCGATTGTCATAGGTCT 699


wt-amt1     TGTGGCTGGAATAGTATGCTACCTTGCTATGGACTTCAGAATAAAGAAGAAGATAGACGA 890
A258T-fwd   TGTGGCTGGAATAGTATGCTACCTTGCTATGGACTTCAGAATAAAGAAGAAGATAGACGA 941
A258T-rev   TGTGGCTGGAATAGTATGCTACCTTGCTATGGACTTCAGAATAAAGAAGAAGATAGACGA 759


wt-amt1     GAGCCTTGATGCTTGGGCGATTCACGGAATAGGCGGTTTATGGGGAAGTGTTGCAGTTGG 950
A258T-fwd   GAGCCTTGATGCTTGGGCGATTCACGGAATAGGCGGTTTATGGGGAAGTGTTGCAGTTGG 1001
A258T-rev   GAGCCTTGATGCTTGGGCGATTCACGGAATAGGCGGTTTATGGGGAAGTGTTGCAGTTGG 819


wt-amt1     CATTCTTGCAAATCCGGA-GGTTAACGGATATGCAGGCCTACTGTTCGGAAATCCGCAAC 1009
A258T-fwd   CATTCTTGCAAATCCGGAAGGTAAACGGATATGCAGGCCTACTGTTCGGAAAT------- 1054
A258T-rev   CATTCTTGCAAATCCGGA-GGTTAACGGATATGCAGGCCTACTGTTCGGAAATCCGCAAC 878


wt-amt1     TGCTAGTTTCACAACTGATTGCGGTTGCATCCACAACAGCCTACGCCTTCCTCGTGACGC 1069
A258T-fwd   ------------------------------------------------------------
A258T-rev   TGCTAGTTTCACAACTGATTGCGGTTGCATCCACAACAGCCTACGCCTTCCTCGTGACGC 938


wt-amt1     TGATACTGGCAAAGGCTGTTGATGCCGCTGTGGGGCTGAGGGTTAGCTCGCAGGAGGAGT 1129
A258T-fwd   ------------------------------------------------------------
A258T-rev   TGATACTGGCAAAGGCTGTTGATGCCGCTGTGGGGCTGAGGGTTAGCTCGCAGGAGGAGT 998


wt-amt1     ACGTCGGTCTCGACCTGTCGCAGCATGAGGAGGTTGCCTACACGTGA------------- 1176
A258T-fwd   ------------------------------------------------------------
A258T-rev   ACGTCGGTCTCGACCTGTCGCAGCATGAGGAGGGGTGCCTACACGCTCGAG<mark style="background:yellow">CACCACCAC</mark> 1058


wt-amt1     --------------------------------------------
A258T-fwd   --------------------------------------------
A258T-rev   <mark style="background:yellow">CACCACCAC</mark>TGAGATCCGGCTGCTAACAAAGCCGAAAGAAGGAGCN 1104
```

D149N

Scores table

```
SeqA Name        Len(nt)  SeqB Name        Len(nt)  Score
=========================================================
1    wt-amt1     1176     2    D149N-fwd   1235     97
1    wt-amt1     1176     3    D149N-rev   1345     98
2    D149N-fwd   1235     3    D149N-rev   1345     94
=========================================================
```

Alignment

```
wt-amt1    ------------------------------------------------------------     
D149N-rev  AAAATTTAATTCGGCTTCCATTTTAGGGGGAATTTGTGAGCGGGATAACCAATTTCCCTT  60
D149N-fwd  -------------------------------------------GTGA--GAATTCCCTC   14


wt-amt1    -----------------------------------------------ATG-AGTGACGG-  11
D149N-rev  TCTAGGAAATAATTTTGGTTTAACTTTAAGGAAGGGAGATTTTCCTTATGGAGTGACGGG  120
D149N-fwd  T-----GAATA--TTTTGTTTA-CTTTAAGAAGGAGATA---TACATATG-AGTGACGG-  61


wt-amt1    AAATGTCG-CATGGATACTCG-CATCCACGGCCCTTGTAA-TGCTGATGGTGCCGGG-AG  67
D149N-rev  AAATGTTGGCATGGATACTTGGCTTCCACGGCCCCTTTAAATGCTGATGGTGCCGGGGAG  180
D149N-fwd  AAATGTCC-CCTGAATACTCG-CATCCACGGCCCTTGTAA-TGCTGATGGTGCCGGG-AG  117


wt-amt1    TGGGGTTCTTTTACGCAGGAATGGTAAGGAGAAAGAATGCAGTTAACATGATTGCGCTGA  127
D149N-rev  TGGGGTTCTTTTACGCAGGAATGGTAAGGAGAAAGAATGCAGTTAACATGATTGCGCTGA  240
D149N-fwd  TGGGGTTCTTTTACGCAGGAATGGTAAGGAGAAAGAATGCAGTTAACATGATTGCGCTGA  177


wt-amt1    GCTTCATATCACTCATAATCACGGTTTTGCTGTGGATATTCTACGGCTACTCGGTGAGCT  187
D149N-rev  GCTTCATATCACTCATAATCACGGTTTTGCTGTGGATATTCTACGGCTACTCGGTGAGCT  300
D149N-fwd  GCTTCATATCACTCATAATCACGGTTTTGCTGTGGATATTCTACGGCTACTCGGTGAGCT  237


wt-amt1    TCGGAAATGACATCTCTGGAATCATTGGAGGGCTGAATTATGCACTGCTAAGCGGAGTTA  247
D149N-rev  TCGGAAATGACATCTCTGGAATCATTGGAGGGCTGAATTATGCACTGCTAAGCGGAGTTA  360
D149N-fwd  TCGGAAATGACATCTCTGGAATCATTGGAGGGCTGAATTATGCACTGCTAAGCGGAGTTA  297


wt-amt1    AGGGGGAGGATTTGCTGTTCATGATGTACCAGATGATGTTCGCCGCTGTCACAATTGCAA  307
D149N-rev  AGGGGGAGGATTTGCTGTTCATGATGTACCAGATGATGTTCGCCGCTGTCACAATTGCAA  420
D149N-fwd  AGGGGGAGGATTTGCTGTTCATGATGTACCAGATGATGTTCGCCGCTGTCACAATTGCAA  357


wt-amt1    TCCTCACCTCCGCAATTGCTGAGAGAGCAAAAGTTTCATCGTTCATTCTCCTCAGCGCTC  367
D149N-rev  TCCTCACCTCCGCAATTGCTGAGAGAGCAAAAGTTTCATCGTTCATTCTCCTCAGCGCTC  480
D149N-fwd  TCCTCACCTCCGCAATTGCTGAGAGAGCAAAAGTTTCATCGTTCATTCTCCTCAGCGCTC  417


wt-amt1    TGTGGCTTACGTTCGTTTACGCCCCCTTCGCACACTGGCTTTGGGGTGGGGGGTGGCTGG  427
D149N-rev  TGTGGCTTACGTTCGTTTACGCCCCCTTCGCACACTGGCTTTGGGGTGGGGGGTGGCTGG  540
D149N-fwd  TGTGGCTTACGTTCGTTTACGCCCCCTTCGCACACTGGCTTTGGGGTGGGGGGTGGCTGG  477


wt-amt1    CAAAGCTCGGCGCCCTCGACTTTGCTGGAGGTATGGTTGTTCACATAAGCTCGGGATTTG  487
D149N-rev  CAAAGCTCGGCGCCCTCAATTTTGCTGGAGGTATGGTTGTTCACATAAGCTCGGGATTTG  600
D149N-fwd  CAAAGCTCGGCGCCCTCAATTTTGCTGGAGGTATGGTTGTTCACATAAGCTCGGGATTTG  537


wt-amt1    CTGCACTTGCAGTCGCGATGACGATAGGTAAGAGGGCGGGATTCGAGGAGTACTCGATAG  547
D149N-rev  CTGCACTTGCAGTCGCGATGACGATAGGTAAGAGGGCGGGATTCGAGGAGTACTCGATAG  660
D149N-fwd  CTGCACTTGCAGTCGCGATGACGATAGGTAAGAGGGCGGGATTCGAGGAGTACTCGATAG  597


wt-amt1    AGCCACACAGCATTCCGCTGACGCTCATTGGCGCTGCCCTGCTTTGGTTTGGGTGGTTCG  607
D149N-rev  AGCCACACAGCATTCCGCTGACGCTCATTGGCGCTGCCCTGCTTTGGTTTGGGTGGTTCG  720
```

```
D149N-fwd    AGCCACACAGCATTCCGCTGACGCTCATTGGCGCTGCCCTGCTTTGGTTTGGGTGGTTCG 657


wt-amt1      GATTCAACGGCGGAAGTGCATTGGCTGCAAACGATGTGGCCATCAACGCCGTGGTGGTCA 667
D149N-rev    GATTCAACGGCGGAAGTGCATTGGCTGCAAACGATGTGGCCATCAACGCCGTGGTGGTCA 780
D149N-fwd    GATTCAACGGCGGAAGTGCATTGGCTGCAAACGATGTGGCCATCAACGCCGTGGTGGTCA 717


wt-amt1      CAAACACCTCAGCAGCAGTAGCAGGGTTTGTCTGGATGGTAATTGGATGGATTAAGGGAA 727
D149N-rev    CAAACACCTCAGCAGCAGTAGCAGGGTTTGTCTGGATGGTAATTGGATGGATTAAGGGAA 840
D149N-fwd    CAAACACCTCAGCAGCAGTAGCAGGGTTTGTCTGGATGGTAATTGGATGGATTAAGGGAA 777


wt-amt1      AGCCGGGGAGTCTTGGGATAGTGAGCGGTGCAATTGCTGGGCTTGCCGCCATAACCCCCG 787
D149N-rev    AGCCGGGGAGTCTTGGGATAGTGAGCGGTGCAATTGCTGGGCTTGCCGCCATAACCCCCG 900
D149N-fwd    AGCCGGGGAGTCTTGGGATAGTGAGCGGTGCAATTGCTGGGCTTGCCGCCATAACCCCCG 837


wt-amt1      CAGCAGGCTTTGTGGATGTAAAGGGAGCGATTGTCATAGGTCTTGTGGCTGGAATAGTAT 847
D149N-rev    CAGCAGGCTTTGTGGATGTAAAGGGAGCGATTGTCATAGGTCTTGTGGCTGGAATAGTAT 960
D149N-fwd    CAGCAGGCTTTGTGGATGTAAAGGGAGCGATTGTCATAGGTCTTGTGGCTGGAATAGTAT 897


wt-amt1      GCTACCTTGCTATGGACTTCAGAATAAAGAAGAAGATAGACGAGAGCCTTGATGCTTGGG 907
D149N-rev    GCTACCTTGCTATGGACTTCAGAATAAAGAAGAAGATAGACGAGAGCCTTGATGCTTGGG 1020
D149N-fwd    GCTACCTTGCTATGGACTTCAGAATAAAGAAGAAGATAGACGAGAGCCTTGATGCTTGGG 957


wt-amt1      CGATTCACGGAATAGGCGGTTTATGGGG-AAGTGTTGCAGTTGGCATTCTTGCAAATCCG 966
D149N-rev    CGATTCACGGAATAGGCGGTTTATGGGG-AAGTGTTGCAGTTGGCATTCTTGCAAATCCG 1079
D149N-fwd    CGATTCACGGAATAGGCGGTTTATGGGGGGAAGTGTTGCAGTTGGCATTCTTGCAAATCCG 1017


wt-amt1      GAGGTTAACGGATATGCAGGCCTACTGTTCGGAAATCCGCAACTGCTAGTTTCACAACTG 1026
D149N-rev    GAGGTTAACGGATATGCAGGCCTACTGTTCGGAAATCCGCAACTGCTAGTTTCACAACTG 1139
D149N-fwd    GAGGTTAACGGATATGCAGGCCTACTGGTCGGAAATCCGCAACTGCTAATTTCACAACTG 1077


wt-amt1      A-TTGCGGTTGCATCCACAACAGCCTAC-GCCTTCCTCGTGA-CGCTGATACTGGC-AAA 1082
D149N-rev    A-TTGCGGTTGCATCCACAACAGCCTAC-GCCTTCCTCGTGA-CGCTGATACTGGC-AAA 1195
D149N-fwd    AATTGGGGTTGCATCACCAACAGCCTACCGCTTTCCTCGTGAACGTTGATACTGGCTAAA 1137


wt-amt1      GGCT-GTTGATGCC-GCTGTGGGGCT--GAGGGTTAGCTCGCAGG--AGGAGTACGTCGG 1136
D149N-rev    GGCT-GTTGATGCC-GCTGTGGGGCT--GAGGGTTAGCTCGCAGG--AGGAGTACGTCGG 1249
D149N-fwd    GGCTTGTTGATGCCCGCTGTGGGGCTTGAAGGGTAACCTCGCAAGGAAGAATTACTTCCG 1197


wt-amt1      --TCTCGACCT-GTCGCAGCATGAGGAGGTTGCCTACACGTGA----------------- 1176
D149N-rev    --TCTCGACCT-GTCGCAGCATGAGGAGGGGGCCTACACGCTCGAGCACCACCACCACCA 1306
D149N-fwd    GTCCTCAACCTTGTCCCAACAT--TGAGGAAGGTTGCCCT------------------- 1235


wt-amt1      ----------------------------------------
D149N-rev    CCACTGAGATCCGGCTGTAACAAGCCGAAAGAGCGGCCC 1345
D149N-fwd    ----------------------------------------
```

H157E

Scores table

```
SeqA Name        Len(nt)  SeqB Name        Len(nt)  Score
=========================================================
1    wt-amt1     1176     2    H157E-fwd   1097     93
1    wt-amt1     1176     3    H157E-rev   1434     98
2    H157E-fwd   1097     3    H157E-rev   1434     96
=========================================================
```

Alignment

```
wt-amt1     ------------------------------------------------------------
H157E-rev   TTTTCCCCTGTGGGTTTGGGGAATATAGCCCCAACAACCCCCCTTGGTGGCCGGGGATCC 60
H157E-fwd   ------------------------------------------------------------


wt-amt1     ------------------------------------------------------------
H157E-rev   CGGGCCAAATGTTTCCCGGGTGAGGGATTGGGATTCTGATCCCCGGAAATTTATTCGCTT 120
H157E-fwd   ------------------------------------------------------------


wt-amt1     ------------------------------------------------------------
H157E-rev   CCCTTTTGGGGGGATTTGGAGCCGGATAACAATTCCCCTTCTAGGAAATAATTTTGTTTA 180
H157E-fwd   --------------------------NTTGCGAATTCCCTCTAAAAAATTTTGTTTAA 32


wt-amt1     --------------------ATGAGTGACGGAAATGTCGCATGGATACTCGCATCCACG 39
H157E-rev   ACTTTAAGAAGGAGATATCATTTGGGGGACGGAAATTTGGCATGGATATTCGCATCCCCG 240
H157E-fwd   CTTTAAGAAGGAGATATACATATGAGTGACGGAAATGTCCCATGGATACTCGCATCCACG 92


wt-amt1     GCCCTTGTAATGCTGATGGTGCCGGGAGTGGGGTTCTTTTACGCAGGAATGGTAAGGAGA 99
H157E-rev   GCCCTTGTAATGCTGATGGTCCCGGGAGTGGGGTTTTTTTACGCAGGAATGGTAAGGGGA 300
H157E-fwd   GCCCTTGTAATGCTGATGGTGCCGGGAGTGGGGTTCTTTTACGCAGGAATGGTAAGGAGA 152


wt-amt1     AAGAATGCAGTTAACATGATTGCGCTGAGCTTCATATCACTCATAATCACGGTTTTGCTG 159
H157E-rev   AAGAATGCAGTTAACATGATTGCGCTGAGCTTCAATTCATTCATAATCACGGTTTTGCTG 360
H157E-fwd   AAGAATGCAGTTAACATGATTGCGCTGAGCTTCATATCACTCATAATCACGGTTTTGCTG 212


wt-amt1     TGGATATTCTACGGCTACTCGGTGAGCTTCGGAAATGACATCTCTGGAATCATTGGAGGG 219
H157E-rev   TGGATATTCTACGGCTACTCGGTGAGCTTCGGAAATGACATCTCTGGAATCATTGGAGGG 420
H157E-fwd   TGGATATTCTACGGCTACTCGGTGAGCTTCGGAAATGACATCTCTGGAATCATTGGAGGG 272


wt-amt1     CTGAATTATGCACTGCTAAGCGGAGTTAAGGGGGAGGATTTGCTGTTCATGATGTACCAG 279
H157E-rev   CTGAATTATGCACTGCTAAGCGGAGTTAAGGGGGAGGATTTGCTGTTCATGATGTACCAG 480
H157E-fwd   CTGAATTATGCACTGCTAAGCGGAGTTAAGGGGGAGGATTTGCTGTTCATGATGTACCAG 332


wt-amt1     ATGATGTTCGCCGCTGTCACAATTGCAATCCTCACCTCCGCAATTGCTGAGAGAGCAAAA 339
H157E-rev   ATGATGTTCGCCGCTGTCACAATTGCAATCCTCACCTCCGCAATTGCTGAGAGAGCAAAA 540
H157E-fwd   ATGATGTTCGCCGCTGTCACAATTGCAATCCTCACCTCCGCAATTGCTGAGAGAGCAAAA 392


wt-amt1     GTTTCATCGTTCATTCTCCTCAGCGCTCTGTGGCTTACGTTCGTTTACGCCCCCTTCGCA 399
H157E-rev   GTTTCATCGTTCATTCTCCTCAGCGCTCTGTGGCTTACGTTCGTTTACGCCCCCTTCGCA 600
H157E-fwd   GTTTCATCGTTCATTCTCCTCAGCGCTCTGTGGCTTACGTTCGTTTACGCCCCCTTCGCA 452


wt-amt1     CACTGGCTTTGGGGTGGGGGGTGGCTGGCAAAGCTCGGCGCCCTCGACTTTGCTGGAGGT 459
H157E-rev   CACTGGCTTTGGGGTGGGGGGTGGCTGGCAAAGCTCGGCGCCCTCGACTTTGCTGGAGGT 660
H157E-fwd   CACTGGCTTTGGGGTGGGGGGTGGCTGGCAAAGCTCGGCGCCCTCGACTTTGCTGGAGGT 512


wt-amt1     ATGGTTGTTCACATAAGCTCGGGATTTGCTGCACTTGCAGTCGCGATGACGATAGGTAAG 519
H157E-rev   ATGGTTGTTGAAATAAGCTCGGGATTTGCTGCACTTGCAGTCGCGATGACGATAGGTAAG 720
H157E-fwd   ATGGTTGTTGAAATAAGCTCGGGATTTGCTGCACTTGCAGTCGCGATGACGATAGGTAAG 572
```

```
wt-amt1     AGGGCGGGATTCGAGGAGTACTCGATAGAGCCACACAGCATTCCGCTGACGCTCATTGGC 579
H157E-rev   AGGGCGGGATTCGAGGAGTACTCGATAGAGCCACACAGCATTCCGCTGACGCTCATTGGC 780
H157E-fwd   AGGGCGGGATTCGAGGAGTACTCGATAGAGCCACACAGCATTCCGCTGACGCTCATTGGC 632


wt-amt1     GCTGCCCTGCTTTGGTTTGGGTGGTTCGGATTCAACGGCGGAAGTGCATTGGCTGCAAAC 639
H157E-rev   GCTGCCCTGCTTTGGTTTGGGTGGTTCGGATTCAACGGCGGAAGTGCATTGGCTGCAAAC 840
H157E-fwd   GCTGCCCTGCTTTGGTTTGGGTGGTTCGGATTCAACGGCGGAAGTGCATTGGCTGCAAAC 692


wt-amt1     GATGTGGCCATCAACGCCGTGGTGGTCACAAACACCTCAGCAGCAGTAGCAGGGTTTGTC 699
H157E-rev   GATGTGGCCATCAACGCCGTGGTGGTCACAAACACCTCAGCAGCAGTAGCAGGGTTTGTC 900
H157E-fwd   GATGTGGCCATCAACGCCGTGGTGGTCACAAACACCTCAGCAGCAGTAGCAGGGTTTGTC 752


wt-amt1     TGGATGGTAATTGGATGGATTAAGGGAAAGCCGGGGAGTCTTGGGATAGTGAGCGGTGCA 759
H157E-rev   TGGATGGTAATTGGATGGATTAAGGGAAAGCCGGGGAGTCTTGGGATAGTGAGCGGTGCA 960
H157E-fwd   TGGATGGTAATTGGATGGATTAAGGGAAAGCCGGGGAGTCTTGGGATAGTGAGCGGTGCA 812


wt-amt1     ATTGCTGGGCTTGCCGCCATAACCCCCGCAGCAGGCTTTGTGGATGTAAAGGGAGCGATT 819
H157E-rev   ATTGCTGGGCTTGCCGCCATAACCCCCGCAGCAGGCTTTGTGGATGTAAAGGGAGCGATT 1020
H157E-fwd   ATTGCTGGGCTTGCCGCCATAACCCCCGCAGCAGGCTTTGTGGATGTAAAGGGAGCGATT 872


wt-amt1     GTCATAGGTCTTGTGGCTGGAATAGTATGCTACCTTGCTATGGACTTCAGAATAAAGAAG 879
H157E-rev   GTCATAGGTCTTGTGGCTGGAATAGTATGCTACCTTGCTATGGACTTCAGAATAAAGAAG 1080
H157E-fwd   GTCATAGGTCTTGTGGCTGGAATAGTATGCTACCTTGCTATGGACTTCAGAATAAAGAAG 932


wt-amt1     AAGATAGACGAGAGCCTTGATGCTTGGG-CGATTCACGG-AATAGGCGG-TTTATGGGGA 936
H157E-rev   AAGATAGACGAGAGCCTTGATGCTTGGG-CGATTCACGG-AATAGGCGG-TTTATGGGGA 1137
H157E-fwd   AAGATAGACGAGAGCCTTGATGCTTGGGGCGATTCACGGGAATAGGCGGGTTTATGGGGA 992


wt-amt1     A-GTGTTGCAGTTGG-CATTCTT-GCAAATCC-GGAGGTT-AACGGA-TATGCA-GGCCT 989
H157E-rev   A-GTGTTGCAGTTGG-CATTCTT-GCAAATCC-GGAGGTT-AACGGA-TATGCA-GGCCT 1190
H157E-fwd   AAGTGTTGCAGTTGGGCATTCTTTGCAAATCCCGGAGGTTTAACGGAATATGCAAGGCCT 1052


wt-amt1     A-CTGTTCGGAAA--TCCGCAA--CTGCTAGTTT-CACAACTGATTGCGGTTGCATCCAC 1043
H157E-rev   A-CTGTTCGGAAA--TCCGCAA--CTGCTAGTTT-CACAACTGATTGCGGTTGCATCCAC 1244
H157E-fwd   AACTGTTCCGGAAATTCCGCCAACTTGCTAATTTTCACAACTTGA--------------- 1097


wt-amt1     AACAGCCTACGCCTTCCTCGTGACGCTGATACTGGCAAAGGCTGTTGATGCCGCTGTGGG 1103
H157E-rev   AACAGCCTACGCCTTCCTCGTGACGCTGATACTGGCAAAGGCTGTTGATGCCGCTGTGGG 1304
H157E-fwd   ------------------------------------------------------------


wt-amt1     GCTGAGGGTTAGCTCGCAGGAGGAGTACGTCGGTCTCGACCTGTCGCAGCATGAGGAGGT 1163
H157E-rev   GCTGAGGGTTAGCTCGCAGGAGGAGTACGTCGGTCTCGACCTGTCGCAGCATGAGGAGTT 1364
H157E-fwd   ------------------------------------------------------------


wt-amt1     TGCCTACACGTGA----------------------------------------------- 1176
H157E-rev   TGCGGACACGCTCGAGCACCACCACCACCACCACTGAGATCCGGCTGTAACAAGCCGAAA 1424
H157E-fwd   ------------------------------------------------------------


wt-amt1     ----------
H157E-rev   GAACGATCCC 1434
H157E-fwd   ----------
```

H157N

Scores table

```
SeqA Name        Len(nt)  SeqB Name        Len(nt)  Score
=========================================================
1    wt-amt1     1176     2    H157N-fwd   1115     92
1    wt-amt1     1176     3    H157N-rev   1419     98
2    H157N-fwd   1115     3    H157N-rev   1419     95
=========================================================
```

Alignment

```
wt-amt1    ------------------------------------------------------------
H157N-rev  GGGGAATCCCGGGCCCCCAGAAGGTTTCCGGCGGTAGAGAGATTCGGATTCTTGGATCCC 60
H157N-fwd  ------------------------------------------------------------


wt-amt1    ------------------------------------------------------------
H157N-rev  CGGGGAAATTTAATTATGATTTCCCTTTTAGGGGGAATTTTGGAGCGGGATAACAAATTC 120
H157N-fwd  ------------------------------------------------------------


wt-amt1    -----------------------------------------------------ATG-AG 5
H157N-rev  CCCTTTCTTGAAAATAAATTTTTGTTTAACTTTTAAGAAAGGGAGGTTATCCATATGGAG 180
H157N-fwd  ---GTGCGAATTCCCTCAGAATATTTTGTTTAACTTTAAGAAGGAGATATACATATG-AG 56


wt-amt1    TGAC--GGAAATGTCGC-ATGGATACT-CGCATCC--ACGGCCC-TTGTAAT-GCTGATG 57
H157N-rev  TGACCGGAAATGTCGCCATGGATACTTCGCATTCCAACGGCCCCTTGTAATTGTTGATG 240
H157N-fwd  TGAC--GGAAATGTCCCCCTAAATACT-CGCATCC--ACGGCCC-TTGTAAT-GCTGATG 109


wt-amt1    GTGCCGGG-AGTGGGGTTCTTTT-ACGCAGGAATGGTAAGGAGAAAGAATGCAGTTAACA 115
H157N-rev  GTGCCGGGGAGTGGGGTTCTTTTTACCCAGGAATGGTAAGGAGAAAGAATGCAGTTAACA 300
H157N-fwd  GTGCCGGG-AGTGGGGTTCTTTT-ACGCAGGAATGGTAAGGAGAAAGAATGCAGTTAACA 167


wt-amt1    TGATTGCGCTGAGCTTCATATCA-CTCATAATCACGGTTTT-GCTGTGGATATTCTACGG 173
H157N-rev  TGATTGCGTTGAGCTTCATATCAATTCATAATCACGGTTTTTGCTGTGGATATTCTACGG 360
H157N-fwd  TGATTGCGCTGAGCTTCATATCA-CTCATAATCACGGTTTT-GCTGTGGATATTCTACGG 225


wt-amt1    CTACTCGGTGAGCTTCGGAAATGACATCTCTGGAATCATTGGAGGGCTGAATTATGCACT 233
H157N-rev  CTACTCGGTGAGCTTCGGAAATGACATCTCTGGAATCATTGGAGGGCTGAATTATGCACT 420
H157N-fwd  CTACTCGGTGAGCTTCGGAAATGACATCTCTGGAATCATTGGAGGGCTGAATTATGCACT 285


wt-amt1    GCTAAGCGGAGTTAAGGGGGAGGATTTGCTGTTCATGATGTACCAGATGATGTTCGCCGC 293
H157N-rev  GCTAAGCGGAGTTAAGGGGGAGGATTTGCTGTTCATGATGTACCAGATGATGTTCGCCGC 480
H157N-fwd  GCTAAGCGGAGTTAAGGGGGAGGATTTGCTGTTCATGATGTACCAGATGATGTTCGCCGC 345


wt-amt1    TGTCACAATTGCAATCCTCACCTCCGCAATTGCTGAGAGAGCAAAAGTTTCATCGTTCAT 353
H157N-rev  TGTCACAATTGCAATCCTCACCTCCGCAATTGCTGAGAGAGCAAAAGTTTCATCGTTCAT 540
H157N-fwd  TGTCACAATTGCAATCCTCACCTCCGCAATTGCTGAGAGAGCAAAAGTTTCATCGTTCAT 405


wt-amt1    TCTCCTCAGCGCTCTGTGGCTTACGTTCGTTTACGCCCCCTTCGCACACTGGCTTTGGGG 413
H157N-rev  TCTCCTCAGCGCTCTGTGGCTTACGTTCGTTTACGCCCCCTTCGCACACTGGCTTTGGGG 600
H157N-fwd  TCTCCTCAGCGCTCTGTGGCTTACGTTCGTTTACGCCCCCTTCGCACACTGGCTTTGGGG 465


wt-amt1    TGGGGGGTGGCTGGCAAAGCTCGGCGCCCTCGACTTTGCTGGAGGTATGGTTGTTCACAT 473
H157N-rev  TGGGGGGTGGCTGGCAAAGCTCGGCGCCCTCGACTTTGCTGGAGGTATGGTTGTTAACAT 660
H157N-fwd  TGGGGGGTGGCTGGCAAAGCTCGGCGCCCTCGACTTTGCTGGAGGTATGGTTGTTAACAT 525


wt-amt1    AAGCTCGGGATTTGCTGCACTTGCAGTCGCGATGACGATAGGTAAGAGGGCGGGATTCGA 533
H157N-rev  AAGCTCGGGATTTGCTGCACTTGCAGTCGCGATGACGATAGGTAAGAGGGCGGGATTCGA 720
H157N-fwd  AAGCTCGGGATTTGCTGCACTTGCAGTCGCGATGACGATAGGTAAGAGGGCGGGATTCGA 585
```

77

```
wt-amt1    GGAGTACTCGATAGAGCCACACAGCATTCCGCTGACGCTCATTGGCGCTGCCCTGCTTTG 593
H157N-rev  GGAGTACTCGATAGAGCCACACAGCATTCCGCTGACGCTCATTGGCGCTGCCCTGCTTTG 780
H157N-fwd  GGAGTACTCGATAGAGCCACACAGCATTCCGCTGACGCTCATTGGCGCTGCCCTGCTTTG 645


wt-amt1    GTTTGGGTGGTTCGGATTCAACGGCGGAAGTGCATTGGCTGCAAACGATGTGGCCATCAA 653
H157N-rev  GTTTGGGTGGTTCGGATTCAACGGCGGAAGTGCATTGGCTGCAAACGATGTGGCCATCAA 840
H157N-fwd  GTTTGGGTGGTTCGGATTCAACGGCGGAAGTGCATTGGCTGCAAACGATGTGGCCATCAA 705


wt-amt1    CGCCGTGGTGGTCACAAACACCTCAGCAGCAGTAGCAGGGTTTGTCTGGATGGTAATTGG 713
H157N-rev  CGCCGTGGTGGTCACAAACACCTCAGCAGCAGTAGCAGGGTTTGTCTGGATGGTAATTGG 900
H157N-fwd  CGCCGTGGTGGTCACAAACACCTCAGCAGCAGTAGCAGGGTTTGTCTGGATGGTAATTGG 765


wt-amt1    ATGGATTAAGGGAAAGCCGGGGAGTCTTGGGATAGTGAGCGGTGCAATTGCTGGGCTTGC 773
H157N-rev  ATGGATTAAGGGAAAGCCGGGGAGTCTTGGGATAGTGAGCGGTGCAATTGCTGGGCTTGC 960
H157N-fwd  ATGGATTAAGGGAAAGCCGGGGAGTCTTGGGATAGTGAGCGGTGCAATTGCTGGGCTTGC 825


wt-amt1    CGCCATAACCCCCGCAGCAGGCTTTGTGGATGTAAAGGGAGCGATTGTCATAGGTCTTGT 833
H157N-rev  CGCCATAACCCCCGCAGCAGGCTTTGTGGATGTAAAGGGAGCGATTGTCATAGGTCTTGT 1020
H157N-fwd  CGCCATAACCCCCGCAGCAGGCTTTGTGGATGTAAAGGGAGCGATTGTCATAGGTCTTGT 885


wt-amt1    GGCTGGAATAGTATGCTACCTTGCTATGGACTTCAGAATAAAGAAGAAGATAGACGAGAG 893
H157N-rev  GGCTGGAATAGTATGCTACCTTGCTATGGACTTCAGAATAAAGAAGAAGATAGACGAGAG 1080
H157N-fwd  GGCTGGAATAGTATGCTACCTTGCTATGGACTTCAGAATAAAGAAGAAGATAGACGAGAG 945


wt-amt1    CCTTGATGCTTGGGCGATTCACGGAATAGGC-GGTTTATGGGGAA-GTGTTGCAGTTGGC 951
H157N-rev  CCTTGATGCTTGGGCGATTCACGGAATAGGC-GGTTTATGGGGAA-GTGTTGCAGTTGGC 1138
H157N-fwd  CCTTGATGCTTGGGCGATTCACGGAATAGGCCGGTTTATGGGGAAAGTGTTGCAGTTGGG 1005


wt-amt1    A-TTCTTG-CAAATCCGGA-GGTTAACGG-ATATGC-AGGCCTACTG-TTCGGAAA--TC 1003
H157N-rev  A-TTCTTG-CAAATCCGGA-GGTTAACGG-ATATGC-AGGCCTACTG-TTCGGAAA--TC 1190
H157N-fwd  AATTCTTGGCAAATCCGGAAGGTTAACGGGATATGCCAGGCCTACTGGTTCGGAAAATCC 1065


wt-amt1    CGCAACT-GCTAGTTT-CACAA-CTGATTGC--GGTTGCATCC-ACAACAGCCTACGCCT 1057
H157N-rev  CGCAACT-GCTAGTTT-CACAA-CTGATTGC--GGTTGCATCC-ACAACAGCCTACGCCT 1244
H157N-fwd  CCCAACTTGCTAAATTACACAAACTGAATGGCGGGTTGCATCCCACAACA---------- 1115


wt-amt1    TCCTCGTGACGCTGATACTGGCAAAGGCTGTTGATGCCGCTGTGGGGCTGAGGGTTAGCT 1117
H157N-rev  TCCTCGTGACGCTGATACTGGCAAAGGCTGTTGATGCCGCTGTGGGGCTGAGGGTTAGCT 1304
H157N-fwd  ------------------------------------------------------------


wt-amt1    CGCAGGAGGAGTACGTCGGTCTCGACCTGTCGCAGCATGAGGAGGTTGCCTACACGTGA- 1176
H157N-rev  CGCAGGAGGAGTACGTCGGTCTCGACCTGTCGCAGCATGAGGAGGGTGCCTACACGCTCG 1364
H157N-fwd  ------------------------------------------------------------


wt-amt1    ------------------------------------------------------------
H157N-rev  AGCACCACCACCACCACCACTGAGATCCGGCTGTAACAAACCCGAAAGAGCGGTC 1419
H157N-fwd  ------------------------------------------------------------
```

I304E

Scores table

```
SeqA Name        Len(nt)  SeqB Name        Len(nt)  Score
==========================================================
1    wt-amt1     1176     2    I304E-fwd   1073     93
1    wt-amt1     1176     3    I304E-rev   1149     91
2    I304E-fwd   1073     3    I304E-rev   1149     82
==========================================================
```

Alignment

```
wt-amt1    ----------------------------------------------------ATGAGT 6
I304E-fwd  CTTGACGAATTCCCTCTAAAAAATTTTGTTTAACTTTAAGAAGGAGATATACATATGAGT 60
I304E-rev  ---------------------------------------------------------- 


wt-amt1    GACGGAAATGTCGCATGGATACTCGCATCCACGGCCCTTGTAATGCTGATGGTGCCGGGA 66
I304E-fwd  GACGGAAATGTCCCATGGATACTCGCATCCACGGCCCTTGTAATGCTGATGGTGCCGGGA 120
I304E-rev  ---------------------------------------------------------- 


wt-amt1    GTGGGGTTCTTTTACGCAGGAATGGTAAGGAGAAAGAATGCAG--TTAACA-TGATT-GC 122
I304E-fwd  GTGGGGTTCTTTTACGCAGGAATGGTAAGGAGAAAGAATGCAG--TTAACA-TGATT-GC 176
I304E-rev  -----------------------------------ATGCCAGTTTAACAATGATTTGC 23


wt-amt1    GCT-GAGCTTC--ATATCACT--CATAATCACGG--TTTTGCTGTGG-ATATTCT-ACGG 173
I304E-fwd  GCT-GAGCTTC--ATATCACT--CATAATCACGG--TTTTGCTGTGG-ATATTCT-ACGG 227
I304E-rev  GCTTGAGCTTTCAATATCACTTCCATAATCACCGGTTTTTGCTGTGGGATATTCTTACGG 83


wt-amt1    -CTACT-CGGTGA-GCTTCGGAAA-TGACATCTCT-GGAATCATTGGAGGGCT-GAATTA 227
I304E-fwd  -CTACT-CGGTGA-GCTTCGGAAA-TGACATCTCT-GGAATCATTGGAGGGCT-GAATTA 281
I304E-rev  GTTACTTCGGTGAAGCTTCGGAAAATGACATTTTTTGGAATCATTGGAAGGCTTGAATTA 143


wt-amt1    TGCACTGCTAAGCGGAGTTAAGGGGGAGGATTTGCTGTTCATGATGTACCAGATGATGTT 287
I304E-fwd  TGCACTGCTAAGCGGAGTTAAGGGGGAGGATTTGCTGTTCATGATGTACCAGATGATGTT 341
I304E-rev  TGCACTTGTAAGCGGAGTTAAGGGGGAGGATTTGCTGTTCATGATGTACCAGATGATGTT 203


wt-amt1    CGCCGCTGTCACAATTGCAATCCTCACCTCCGCAATTGCTGAGAGAGCAAAAGTTTCATC 347
I304E-fwd  CGCCGCTGTCACAATTGCAATCCTCACCTCCGCAATTGCTGAGAGAGCAAAAGTTTCATC 401
I304E-rev  CGCCGCTGTCACAATTGCAATCCTCACCTCCGCAATTGCTGAGAGAGCAAAAGTTTCATC 263


wt-amt1    GTTCATTCTCCTCAGCGCTCTGTGGCTTACGTTCGTTTACGCCCCCTTCGCACACTGGCT 407
I304E-fwd  GTTCATTCTCCTCAGCGCTCTGTGGCTTACGTTCGTTTACGCCCCCTTCGCACACTGGCT 461
I304E-rev  GTTCATTCTCCTCAGCGCTCTGTGGCTTACGTTCGTTTACGCCCCCTTCGCACACTGGCT 323


wt-amt1    TTGGGGTGGGGGGTGGCTGGCAAAGCTCGGCGCCCTCGACTTTGCTGGAGGTATGGTTGT 467
I304E-fwd  TTGGGGTGGGGGGTGGCTGGCAAAGCTCGGCGCCCTCGACTTTGCTGGAGGTATGGTTGT 521
I304E-rev  TTGGGGTGGGGGGTGGCTGGCAAAGCTCGGCGCCCTCGACTTTGCTGGAGGTATGGTTGT 383


wt-amt1    TCACATAAGCTCGGGATTTGCTGCACTTGCAGTCGCGATGACGATAGGTAAGAGGGCGGG 527
I304E-fwd  TCACATAAGCTCGGGATTTGCTGCACTTGCAGTCGCGATGACGATAGGTAAGAGGGCGGG 581
I304E-rev  TCACATAAGCTCGGGATTTGCTGCACTTGCAGTCGCGATGACGATAGGTAAGAGGGCGGG 443


wt-amt1    ATTCGAGGAGTACTCGATAGAGCCACACAGCATTCCGCTGACGCTCATTGGCGCTGCCCT 587
I304E-fwd  ATTCGAGGAGTACTCGATAGAGCCACACAGCATTCCGCTGACGCTCATTGGCGCTGCCCT 641
I304E-rev  ATTCGAGGAGTACTCGATAGAGCCACACAGCATTCCGCTGACGCTCATTGGCGCTGCCCT 503


wt-amt1    GCTTTGGTTTGGGTGGTTCGGATTCAACGGCGGAAGTGCATTGGCTGCAAACGATGTGGC 647
I304E-fwd  GCTTTGGTTTGGGTGGTTCGGATTCAACGGCGGAAGTGCATTGGCTGCAAACGATGTGGC 701
I304E-rev  GCTTTGGTTTGGGTGGTTCGGATTCAACGGCGGAAGTGCATTGGCTGCAAACGATGTGGC 563
```

```
wt-amt1    CATCAACGCCGTGGTGGTCACAAACACCTCAGCAGCAGTAGCAGGGTTTGTCTGGATGGT 707
I304E-fwd  CATCAACGCCGTGGTGGTCACAAACACCTCAGCAGCAGTAGCAGGGTTTGTCTGGATGGT 761
I304E-rev  CATCAACGCCGTGGTGGTCACAAACACCTCAGCAGCAGTAGCAGGGTTTGTCTGGATGGT 623


wt-amt1    AATTGGATGGATTAAGGGAAAGCCGGGGAGTCTTGGGATAGTGAGCGGTGCAATTGCTGG 767
I304E-fwd  AATTGGATGGATTAAGGGAAAGCCGGGGAGTCTTGGGATAGTGAGCGGTGCAATTGCTGG 821
I304E-rev  AATTGGATGGATTAAGGGAAAGCCGGGGAGTCTTGGGATAGTGAGCGGTGCAATTGCTGG 683


wt-amt1    GCTTGCCGCCATAACCCCCGCAGCAGGCTTTGTGGATGTAAAGGGAGCGATTGTCATAGG 827
I304E-fwd  GCTTGCCGCCATAACCCCCGCAGCAGGCTTTGTGGATGTAAAGGGAGCGATTGTCATAGG 881
I304E-rev  GCTTGCCGCCATAACCCCCGCAGCAGGCTTTGTGGATGTAAAGGGAGCGATTGTCATAGG 743


wt-amt1    TCTTGTGGCTGGAATAGTATGCTACCTTGCTATGGACTTCAGAATAAAGAAGAAGATAGA 887
I304E-fwd  TCTTGTGGCTGGAATAGTATGCTACCTTGCTATGGACTTCAGAATAAAGAAGAAGATAGA 941
I304E-rev  TCTTGTGGCTGGAATAGTATGCTACCTTGCTATGGACTTCAGAATAAAGAAGAAGATAGA 803


wt-amt1    CGAGAGCCTTGATGCTTGGGCGATTCACGGAATA-GGCGGTTTATGGGGAA-GTGTTGCA 945
I304E-fwd  CGAGAGCCTTGATGCTTGGGCGGAACACGGAATAAGGCGGTTTATGGGGAAAGTGTTGCA 1001
I304E-rev  CGAGAGCCTTGATGCTTGGGCGGAACACGGAATA-GGCGGTTTATGGGGAA-GTGTTGCA 861


wt-amt1    GTT-GGCATTCTTGCAAATCC-GGAGGTTAACGGATAT-GCAGGCC-TACTGTTCGGAAA 1001
I304E-fwd  ATTTGGCTTTCTTGCAAATCCTGAAGGTTAACGGAAATTGCAGGCCCTACTGTTCCGGAA 1061
I304E-rev  GTT-GGCATTCTTGCAAATCC-GGAGGTTAACGGATAT-GCAGGCC-TACTGTTCGGAAA 917


wt-amt1    TCCGCAACTGCTAGTTTCACAACTGATTGCGGTTGCATCCACAACAGCCTACGCCTTCCT 1061
I304E-fwd  AATCCCCCAACT------------------------------------------------ 1073
I304E-rev  TCCGCAACTGCTAGTTTCACAACTGATTGCGGTTGCATCCACAACAGCCTACGCCTTCCT 977


wt-amt1    CGTGACGCTGATACTGGCAAAGGCTGTTGATGCCGCTGTGGGGCTGAGGGTTAGCTCGCA 1121
I304E-fwd  ------------------------------------------------------------
I304E-rev  CGTGACGCTGATACTGGCAAAGGCTGTTGATGCCGCTGTGGGGCTGAGGGTTAGCTCGCA 1037


wt-amt1    GGAGGAGTACGTCGGTCTCGACCTGTCGCAGCATGAGGAGGTTGCCTACACGTGA----- 1176
I304E-fwd  ------------------------------------------------------------
I304E-rev  GGAGGAGTACGTCGGTCTCGACCTGTCGCAGCATGAGGAGGTTGCCTACACGCTCGAGCA 1097


wt-amt1    ---------------------------------------------------
I304E-fwd  ---------------------------------------------------
I304E-rev  CCACCACCACCACCACTGAGATCCGGCTGCTAACAAACCGAAAGAACGGCCC 1149
```

N205L

Scores table

```
SeqA Name        Len(nt)  SeqB Name        Len(nt)  Score
=========================================================
1    wt-amt1     1176     2    N205L-fwd    878      88
=========================================================
```

Alignment

```
wt-amt1    ------------------------------------------------------------
N205L-fwd  GGTGAAGGAACCTCTCCCCTCCTATAATAATTTTGTTTAACTTTAAGAAGGAGATATACA 60


wt-amt1    -ATGAGTGACGGAAATGTCGCATGGATACTCGCATCCACGGCCCTTGTAATGCTGATGGT 59
N205L-fwd  TATGAGTGACGGAAATGT-GGATGGGTACTCGCATCCACGGCCCTTGTAATGCTGATGGT 119


wt-amt1    GCCGGGAGTGGGGTTCTTTTACGCAGGAATGGTAAGGAGAAAGAATGCAGTTAACATGAT 119
N205L-fwd  GCCGGGAGTGGGGTTCTTTTACGCAGGAATGGTAAGGAGAAAGAATGCAGTTAACATGAT 179


wt-amt1    TGCGCTGAGCTTCATATCACTCATAATCACGGTTTTGCTGTGGATATTCTACGGCTACTC 179
N205L-fwd  TGCGCTGAGCTTCATATCACTCATAATCACGGTTTTGCTGTGGATATTCTACGGCTACTC 239


wt-amt1    GGTGAGCTTCGGAAATGACATCTCTGGAATCATTGGAGGGCTGAATTATGCACTGCTAAG 239
N205L-fwd  GGTGAGCTTCGGAAATGACATCTCTGGAATCATTGGAGGGCTGAATTATGCACTGCTAAG 299


wt-amt1    CGGAGTTAAGGGGGAGGATTTGCTGTTCATGATGTACCAGATGATGTTCGCCGCTGTCAC 299
N205L-fwd  CGGAGTTAAGGGGGAGGATTTGCTGTTCATGATGTACCAGATGATGTTCGCCGCTGTCAC 359


wt-amt1    AATTGCAATCCTCACCTCCGCAATTGCTGAGAGAGCAAAAGTTTCATCGTTCATTCTCCT 359
N205L-fwd  AATTGCAATCCTCACCTCCGCAATTGCTGAGAGAGCAAAAGTTTCATCGTTCATTCTCCT 419


wt-amt1    CAGCGCTCTGTGGCTTACGTTCGTTTACGCCCCCTTCGCACACTGGCTTTGGGGTGGGGG 419
N205L-fwd  CAGCGCTCTGTGGCTTACGTTCGTTTACGCCCCCTTCGCACACTGGCTTTGGGGTGGGGG 479


wt-amt1    GTGGCTGGCAAAGCTCGGCGCCCTCGACTTTGCTGGAGGTATGGTTGTTCACATAAGCTC 479
N205L-fwd  -TGGCTGGCAAAGCTCGGCGACCTCGACTTTGCTGGAGGCATGGTTGTTCACATAACCTC 538


wt-amt1    GGGATTTGCTGCACTTGCAGTCGCGATGACGATAGGTAAGAGGGCGGGATTCGAGGAGTA 539
N205L-fwd  GGGATTTGCTGCACTTGCCGTCGCGATGACGATAGGTAATAGGGCGGGATTCGAGGAGTA 598


wt-amt1    CTCGATAGAGCCACACAGCATTCCGCTGACGCTCATTGGCGCTGCCCTGCTTTGGTTTGG 599
N205L-fwd  CTCGATAGAGCCACACAGCATTCCGCTGACGCTCACTGGCGCTGCCCTGCACTGGTTTGG 658


wt-amt1    GTGGTTCGGATTCAACGGCGGAAGTGCATTGGCTGCAAACGATGTGGCCATCAACGCCGT 659
N205L-fwd  GTGGTTCGTATTCCTTGGCGGAAGAGCCTTGTCTGCAAACGATGTGGCCATCAACGCCGT 718


wt-amt1    GGTGGTCACAAACACCTCAGCAGCAGTAGCAGGGTTTGTCTGGATGGTAATTGGATGGAT 719
N205L-fwd  GGTGGTCACAAACACCTCAGC--TCATAACAGGGTTTGTCTGGATGGAAATTGAATGA-- 774


wt-amt1    TAAGGGAAAGCCGGGGGAGTCTTGGGATAGTGAGCGGTGCAATTGCTGGGCTTGCCGCCAT 779
N205L-fwd  TAATGCAAAGTCCGG-AGTCATGG-ATAATGAGCGCTG-AATTGCTGGGCTT-TCACCTT 830


wt-amt1    AACCCCCGCAGCAGGCTTTGTGGATGTAAAGGGAGCGATTGTCTATAGGTCTTGTGGCTGG 839
N205L-fwd  A-CCCCAGCAGCCCGTTTTTGGGATGTAAAGG-AACGATTGTTATAGGTT---------- 878


wt-amt1    AATAGTATGCTACCTTGCTATGGACTTCAGAATAAAGAAGAAGATAGACGAGAGCCTTGA 899
N205L-fwd  ------------------------------------------------------------
```

```
wt-amt1      TGCTTGGGCGATTCACGGAATAGGCGGTTTATGGGGAAGTGTTGCAGTTGGCATTCTTGC 959
N205L-fwd    ------------------------------------------------------------


wt-amt1      AAATCCGGAGGTTAACGGATATGCAGGCCTACTGTTCGGAAATCCGCAACTGCTAGTTTC 1019
N205L-fwd    ------------------------------------------------------------


wt-amt1      ACAACTGATTGCGGTTGCATCCACAACAGCCTACGCCTTCCTCGTGACGCTGATACTGGC 1079
N205L-fwd    ------------------------------------------------------------


wt-amt1      AAAGGCTGTTGATGCCGCTGTGGGGCTGAGGGTTAGCTCGCAGGAGGAGTACGTCGGTCT 1139
N205L-fwd    ------------------------------------------------------------


wt-amt1      CGACCTGTCGCAGCATGAGGAGGTTGCCTACACGTGA 1176
N205L-fwd    -------------------------------------
```

T261V

Scores table

SeqA	Name	Len(nt)	SeqB	Name	Len(nt)	Score
1	wt-amt1	1176	2	T261V-fwd	1228	98
1	wt-amt1	1176	3	T261V-rev	1407	99
2	T261V-fwd	1228	3	T261V-rev	1407	96

Alignment

```
wt-amt1      ------------------------------------------------------------
T261V-rev    CAAAGAATTGTTTCCCGCGGTTAAAGAGATTCGGGATTTTTCGATTCCCCGGGAAAATTA 60
T261V-fwd    ------------------------------------------------------------


wt-amt1      ------------------------------------------------------------
T261V-rev    ATTCCGACCTCGCTTTTAGGGGGAATTTGTGAGGCGGGATAACCAATTTCCCCTTTTAAG 120
T261V-fwd    ----------------------------------------------------GTGAG 5


wt-amt1      ------------------------------------------ATGAGTGAC--GGAA 13
T261V-rev    AAATTAATTTTGGTTTTAACTTTTAAGAAGGGGAGGATTTTCCATTATGAGTGGACGGGAA 180
T261V-fwd    AATTCCCTCTGAATATTTTGTTTACTTTAAGAAGGAGATATACATATGAGTGAC--GGAA 63


wt-amt1      ATGTCGCA-TGGATACTCG-CATCCA-CGGCCC-TTGTAATGCTGATGGTGCCGGG-AGT 68
T261V-rev    ATGTCGCAATGGATACTTGGCATCCAACGGCCCCTTGTAATGCTGATGGTGCCGGGGAGT 240
T261V-fwd    ATGTCCCA-TGGATACTCG-CATCCA-CGGCCC-TTGTAATGCTGATGGTGCCGGG-AGT 118


wt-amt1      GGGGTT-CTTTTACGC-AGGAATGGTAAGGAGAAAGAATGCAGTTAACATGATTGCGCTG 126
T261V-rev    GGGGTTTCTTTTACGCCAGGAATGGTAAGGAGAAAGAATGCAGTTAACATGATTGCGCTG 300
T261V-fwd    GGGGTT-CTTTTACGC-AGGAATGGTAAGGAGAAAGAATGCAGTTAACATGATTGCGCTG 176


wt-amt1      AGCTTCATATCACTCATAATCACGGTTTTGCTGTGGATATTCTACGGCTACTCGGTGAGC 186
T261V-rev    AGCTTCATATCACTCATAATCACGGTTTTGCTGTGGATATTCTACGGCTACTCGGTGAGC 360
T261V-fwd    AGCTTCATATCACTCATAATCACGGTTTTGCTGTGGATATTCTACGGCTACTCGGTGAGC 236


wt-amt1      TTCGGAAATGACATCTCTGGAATCATTGGAGGGCTGAATTATGCACTGCTAAGCGGAGTT 246
T261V-rev    TTCGGAAATGACATCTCTGGAATCATTGGAGGGCTGAATTATGCACTGCTAAGCGGAGTT 420
T261V-fwd    TTCGGAAATGACATCTCTGGAATCATTGGAGGGCTGAATTATGCACTGCTAAGCGGAGTT 296


wt-amt1      AAGGGGGAGGATTTGCTGTTCATGATGTACCAGATGATGTTCGCCGCTGTCACAATTGCA 306
T261V-rev    AAGGGGGAGGATTTGCTGTTCATGATGTACCAGATGATGTTCGCCGCTGTCACAATTGCA 480
```

```
T261V-fwd    AAGGGGGAGGATTTGCTGTTCATGATGTACCAGATGATGTTCGCCGCTGTCACAATTGCA 356


wt-amt1      ATCCTCACCTCCGCAATTGCTGAGAGAGCAAAAGTTTCATCGTTCATTCTCCTCAGCGCT 366
T261V-rev    ATCCTCACCTCCGCAATTGCTGAGAGAGCAAAAGTTTCATCGTTCATTCTCCTCAGCGCT 540
T261V-fwd    ATCCTCACCTCCGCAATTGCTGAGAGAGCAAAAGTTTCATCGTTCATTCTCCTCAGCGCT 416


wt-amt1      CTGTGGCTTACGTTCGTTTACGCCCCCTTCGCACACTGGCTTTGGGGTGGGGGGTGGCTG 426
T261V-rev    CTGTGGCTTACGTTCGTTTACGCCCCCTTCGCACACTGGCTTTGGGGTGGGGGGTGGCTG 600
T261V-fwd    CTGTGGCTTACGTTCGTTTACGCCCCCTTCGCACACTGGCTTTGGGGTGGGGGGTGGCTG 476


wt-amt1      GCAAAGCTCGGCGCCCTCGACTTTGCTGGAGGTATGGTTGTTCACATAAGCTCGGGATTT 486
T261V-rev    GCAAAGCTCGGCGCCCTCGACTTTGCTGGAGGTATGGTTGTTCACATAAGCTCGGGATTT 660
T261V-fwd    GCAAAGCTCGGCGCCCTCGACTTTGCTGGAGGTATGGTTGTTCACATAAGCTCGGGATTT 536


wt-amt1      GCTGCACTTGCAGTCGCGATGACGATAGGTAAGAGGGCGGGATTCGAGGAGTACTCGATA 546
T261V-rev    GCTGCACTTGCAGTCGCGATGACGATAGGTAAGAGGGCGGGATTCGAGGAGTACTCGATA 720
T261V-fwd    GCTGCACTTGCAGTCGCGATGACGATAGGTAAGAGGGCGGGATTCGAGGAGTACTCGATA 596


wt-amt1      GAGCCACACAGCATTCCGCTGACGCTCATTGGCGCTGCCCTGCTTTGGTTTGGGTGGTTC 606
T261V-rev    GAGCCACACAGCATTCCGCTGACGCTCATTGGCGCTGCCCTGCTTTGGTTTGGGTGGTTC 780
T261V-fwd    GAGCCACACAGCATTCCGCTGACGCTCATTGGCGCTGCCCTGCTTTGGTTTGGGTGGTTC 656


wt-amt1      GGATTCAACGGCGGAAGTGCATTGGCTGCAAACGATGTGGCCATCAACGCCGTGGTGGTC 666
T261V-rev    GGATTCAACGGCGGAAGTGCATTGGCTGCAAACGATGTGGCCATCAACGCCGTGGTGGTC 840
T261V-fwd    GGATTCAACGGCGGAAGTGCATTGGCTGCAAACGATGTGGCCATCAACGCCGTGGTGGTC 716


wt-amt1      ACAAACACCTCAGCAGCAGTAGCAGGGTTTGTCTGGATGGTAATTGGATGGATTAAGGGA 726
T261V-rev    ACAAACACCTCAGCAGCAGTAGCAGGGTTTGTCTGGATGGTAATTGGATGGATTAAGGGA 900
T261V-fwd    ACAAACACCTCAGCAGCAGTAGCAGGGTTTGTCTGGATGGTAATTGGATGGATTAAGGGA 776


wt-amt1      AAGCCGGGGAGTCTTGGGATAGTGAGCGGTGCAATTGCTGGGCTTGCCGCCATAACCCCC 786
T261V-rev    AAGCCGGGGAGTCTTGGGATAGTGAGCGGTGCAATTGCTGGGCTTGCCGCCATA[GTC]CCC 960
T261V-fwd    AAGCCGGGGAGTCTTGGGATAGTGAGCGGTGCAATTGCTGGGCTTGCCGCCATA[GTC]CCC 836


wt-amt1      GCAGCAGGCTTTGTGGATGTAAAGGGAGCGATTGTCATAGGTCTTGTGGCTGGAATAGTA 846
T261V-rev    GCAGCAGGCTTTGTGGATGTAAAGGGAGCGATTGTCATAGGTCTTGTGGCTGGAATAGTA 1020
T261V-fwd    GCAGCAGGCTTTGTGGATGTAAAGGGAGCGATTGTCATAGGTCTTGTGGCTGGAATAGTA 896


wt-amt1      TGCTACCTTGCTATGGACTTCAGAATAAAGAAGAAGATAGACGAGAGCCTTGATGCTTGG 906
T261V-rev    TGCTACCTTGCTATGGACTTCAGAATAAAGAAGAAGATAGACGAGAGCCTTGATGCTTGG 1080
T261V-fwd    TGCTACCTTGCTATGGACTTCAGAATAAAGAAGAAGATAGACGAGAGCCTTGATGCTTGG 956


wt-amt1      GCGATTCACGGAATAGGCGGTTTATGGGGAAGTGTTGCAGTTGGCATTCTTGCAAATCCG 966
T261V-rev    GCGATTCACGGAATAGGCGGTTTATGGGGAAGTGTTGCAGTTGGCATTCTTGCAAATCCG 1140
T261V-fwd    GCGATTCACGGAATAGGCGGTTTATGGGGAAGTGTTGCAGTTGGCATTCTTGCAAATCCG 1016


wt-amt1      GAGGTTAACGGATATGCAGGCCTACTGTTCGGAAATCCGCAACTGCTAGTTTCACAACTG 1026
T261V-rev    GAGGTTAACGGATATGCAGGCCTACTGTTCGGAAATCCGCAACTGCTAGTTTCACAACTG 1200
T261V-fwd    GAGGTTAACGGATATGCAGGCCTACTGTTCGGAAATCCGCAACTGCTAGTTTCACAACTG 1076


wt-amt1      ATTGCGGTTGCATCCACAACAGCCTACGCCTTCCTCGTGACGCTGATACTGGCAAAGGCT 1086
T261V-rev    ATTGCGGTTGCATCCACAACAGCCTACGCCTTCCTCGTGACGCTGATACTGGCAAAGGCT 1260
T261V-fwd    ATTGCGGTTGCATCCACAACAGCCTACGCCTTCCTCGTGACGCTGATACTGGCAAAAGCT 1136


wt-amt1      -GTTGATGCCGCT-GTGGGGCTGA-GGGTTAGCTCGCAGGA-GGAGTACGTC-GGTCTCG 1141
T261V-rev    -GTTGATGCCGCT-GTGGGGCTGA-GGGTTAGCTCGCAGGA-GGAGTACGTC-GGTCTCG 1315
T261V-fwd    TGTTGATGCCGCTTGTGGGGCTGAAGGGTAAGCTCGCAAGAAGGAGTACGTCAGGTCTCA 1196


wt-amt1      ACCTGTCGCAGCATGAGGAGGTTGCCTACACGTGA------------------------- 1176
T261V-rev    ACCTGTCGCAGCATGAGGAGGTTGCCTACACGCTCGAG[CACCACCACCACCACCAC]TGAG 1375
T261V-fwd    ACCTGCCCCACAATGAGGAAGTTGCTCA-ACAN------------------------- 1228
```

83

```
wt-amt1         --------------------------------
T261V-rev       ATCCGGCTGTAACAAGCCCGAAAGAGCGGCCC 1407
T261V-fwd       --------------------------------
```

V156Δ

Scores table

```
SeqA Name               Len(nt)  SeqB Name               Len(nt)  Score
=======================================================================
1    wt-amt1            1176     2    V156delta-fwd      1022     93
1    wt-amt1            1176     3    V156delta-rev      1452     98
2    V156delta-fwd      1022     3    V156delta-rev      1452     95
=======================================================================
```

Alignment

```
wt-amt1         ------------------------------------------------------------
V156delta-rev   CGGGGGATTAAGGGCCCCAAGAAAACCGCCACTTTGGGGGCTCCCGGTAATCCCGGCCCC 60
V156delta-fwd   ------------------------------------------------------------


wt-amt1         ------------------------------------------------------------
V156delta-rev   AACAAAGGGTTCCGGCGTTTGAAGGTATGGAGGATTCTTGATTCCCCGCGGAAAATTAAT 120
V156delta-fwd   ------------------------------------------------------------


wt-amt1         ------------------------------------------------------------
V156delta-rev   TACGGACTTCACTTTAAGGGGGAATTTGGGAGCGGGAATACAATTTCCCCTTCTAGGAAA 180
V156delta-fwd   ------------------------------------------------GTGACGAATTCCCTC 15


wt-amt1         -----------------------------------ATG-AGTGACGGAAA-TGTCGC- 20
V156delta-rev   TAATTTGTGTTTAACTTTAAGAAAGGAGGTTTACCATATGGAGTGACGGAAAATGTTGCC 240
V156delta-fwd   AGAATATTTTGTTTAACTTTAAGAAGGAGATATACATATG-AGTGACGGAAA-TGTCCC- 72


wt-amt1         ATGGATACT-CGCATCCACGG--CCCTTGTAATGCTGATGGTGCCGGG-AGTGGGGTTCT 76
V156delta-rev   ATGGATACTTCGCATTCCCCGGCCCCTTGTAATGCTGATGGTGCCGGGGAGTGGGGTTCT 300
V156delta-fwd   ATGGATACT-CGCATCCACGG--CCCTTGTAATGCTGATGGTGCCGGG-AGTGGGGTTCT 128


wt-amt1         TTTACGCAGGAATGGTAAGGAGAAAGAATGCAGTTAACATGATTGCGCTGAGCTTCATAT 136
V156delta-rev   TTACGCCAGGAATGGTAAGGAGAAAGAATGCAGTTAACATGATTGCGCTGAGCTTCATAT 360
V156delta-fwd   TTTACGCAGGAATGGTAAGGAGAAAGAATGCAGTTAACATGATTGCGCTGAGCTTCATAT 188


wt-amt1         CACTCATAATCACGGTTTTGCTGTGGATATTCTACGGCTACTCGGTGAGCTTCGGAAATG 196
V156delta-rev   CACTCATAATCACGGTTTTGCTGTGGATATTCTACGGCTACTCGGTGAGCTTCGGAAATG 420
V156delta-fwd   CACTCATAATCACGGTTTTGCTGTGGATATTCTACGGCTACTCGGTGAGCTTCGGAAATG 248


wt-amt1         ACATCTCTGGAATCATTGGAGGGCTGAATTATGCACTGCTAAGCGGAGTTAAGGGGGAGG 256
V156delta-rev   ACATCTCTGGAATCATTGGAGGGCTGAATTATGCACTGCTAAGCGGAGTTAAGGGGGAGG 480
V156delta-fwd   ACATCTCTGGAATCATTGGAGGGCTGAATTATGCACTGCTAAGCGGAGTTAAGGGGGAGG 308


wt-amt1         ATTTGCTGTTCATGATGTACCAGATGATGTTCGCCGCTGTCACAATTGCAATCCTCACCT 316
V156delta-rev   ATTTGCTGTTCATGATGTACCAGATGATGTTCGCCGCTGTCACAATTGCAATCCTCACCT 540
V156delta-fwd   ATTTGCTGTTCATGATGTACCAGATGATGTTCGCCGCTGTCACAATTGCAATCCTCACCT 368


wt-amt1         CCGCAATTGCTGAGAGAGCAAAAGTTTCATCGTTCATTCTCCTCAGCGCTCTGTGGCTTA 376
V156delta-rev   CCGCAATTGCTGAGAGAGCAAAAGTTTCATCGTTCATTCTCCTCAGCGCTCTGTGGCTTA 600
V156delta-fwd   CCGCAATTGCTGAGAGAGCAAAAGTTTCATCGTTCATTCTCCTCAGCGCTCTGTGGCTTA 428


wt-amt1         CGTTCGTTTACGCCCCCTTCGCACACTGGCTTTGGGGTGGGGGGTGGCTGGCAAAGCTCG 436
V156delta-rev   CGTTCGTTTACGCCCCCTTCGCACACTGGCTTTGGGGTGGGGGGTGGCTGGCAAAGCTCG 660
V156delta-fwd   CGTTCGTTTACGCCCCCTTCGCACACTGGCTTTGGGGTGGGGGGTGGCTGGCAAAGCTCG 488
```

```
wt-amt1        GCGCCCTCGACTTTGCTGGAGGTATGGTTGTTCACATAAGCTCGGGATTTGCTGCACTTG 496
V156delta-rev  GCGCCCTCGACTTTGCTGGAGGTATGGTT---CACATAAGCTCGGGATTTGCTGCACTTG 717
V156delta-fwd  GCGCCCTCGACTTTGCTGGAGGTATGGTT---CACATAAGCTCGGGATTTGCTGCACTTG 545


wt-amt1        CAGTCGCGATGACGATAGGTAAGAGGGCGGGATTCGAGGAGTACTCGATAGAGCCACACA 556
V156delta-rev  CAGTCGCGATGACGATAGGTAAGAGGGCGGGATTCGAGGAGTACTCGATAGAGCCACACA 777
V156delta-fwd  CAGTCGCGATGACGATAGGTAAGAGGGCGGGATTCGAGGAGTACTCGATAGAGCCACACA 605


wt-amt1        GCATTCCGCTGACGCTCATTGGCGCTGCCCTGCTTTGGTTTGGGTGGTTCGGATTCAACG 616
V156delta-rev  GCATTCCGCTGACGCTCATTGGCGCTGCCCTGCTTTGGTTTGGGTGGTTCGGATTCAACG 837
V156delta-fwd  GCATTCCGCTGACGCTCATTGGCGCTGCCCTGCTTTGGTTTGGGTGGTTCGGATTCAACG 665


wt-amt1        GCGGAAGTGCATTGGCTGCAAACGATGTGGCCATCAACGCCGTGGTGGTCACAAACACCT 676
V156delta-rev  GCGGAAGTGCATTGGCTGCAAACGATGTGGCCATCAACGCCGTGGTGGTCACAAACACCT 897
V156delta-fwd  GCGGAAGTGCATTGGCTGCAAACGATGTGGCCATCAACGCCGTGGTGGTCACAAACACCT 725


wt-amt1        CAGCAGCAGTAGCAGGGTTTGTCTGGATGGTAATTGGATGGATTAAGGGAAAGCCGGGGA 736
V156delta-rev  CAGCAGCAGTAGCAGGGTTTGTCTGGATGGTAATTGGATGGATTAAGGGAAAGCCGGGGA 957
V156delta-fwd  CAGCAGCAGTAGCAGGGTTTGTCTGGATGGTAATTGGATGGATTAAGGGAAAGCCGGGGA 785


wt-amt1        GTCTTGGGATAGTGAGCGGTGCAATTGCTGGGCTTGCCGCCATAACCCCCGCAGCAGGCT 796
V156delta-rev  GTCTTGGGATAGTGAGCGGTGCAATTGCTGGGCTTGCCGCCATAACCCCCGCAGCAGGCT 1017
V156delta-fwd  GTCTTGGGATAGTGAGCGGTGCAATTGCTGGGCTTGCCGCCATAACCCCCGCAGCAGGCT 845


wt-amt1        TTGTGGATGTAAAGGGAGCGATTGTCATAGGTCTTGTGGCTGGAATAGTATGCTACCTTG 856
V156delta-rev  TTGTGGATGTAAAGGGAGCGATTGTCATAGGTCTTGTGGCTGGAATAGTATGCTACCTTG 1077
V156delta-fwd  TTGTGGATGTAAAGGGAGCGATTGTCATAGGTCTTGTGGCTGGAATAGTATGCTACCTTG 905


wt-amt1        CTATGGACTTCAGAATAAAGAAGAA-GATAGACGAGA-GCCTTGATGCTTGGGC-GATTC 913
V156delta-rev  CTATGGACTTCAGAATAAAGAAGAA-GATAGACGAGA-GCCTTGATGCTTGGGC-GATTC 1134
V156delta-fwd  CTATGGACTTCAGAATAAAGAAGAAAGATAGACGAGAAGCCTTGATGCTTGGGCCGATTC 965


wt-amt1        ACGGAATA-GGCGGTTTATGGGGAA-GTGTT-GCAGTTG-GCATT-CTTGCAAATCCGGA 968
V156delta-rev  ACGGAATA-GGCGGTTTATGGGGAA-GTGTT-GCAGTTG-GCATT-CTTGCAAATCCGGA 1189
V156delta-fwd  ACGGAATAAGGTGGTTTATGGGGAAAGTGTTTGTAGTTGTGTATTATTTGGAAATGA--- 1022


wt-amt1        GGTTAACGGATATGCAGGCCTACTGTTCGGAAATCCGCAACTGCTAGTTTCACAACTGAT 1028
V156delta-rev  GGTTAACGGATATGCAGGCCTACTGTTCGGAAATCCGCAACTGCTAGTTTCACAACTGAT 1249
V156delta-fwd  ------------------------------------------------------------


wt-amt1        TGCGGTTGCATCCACAACAGCCTACGCCTTCCTCGTGACGCTGATACTGGCAAAGGCTGT 1088
V156delta-rev  TGCGGTTGCATCCACAACAGCCTACGCCTTCCTCGTGACGCTGATACTGGCAAAGGCTGT 1309
V156delta-fwd  ------------------------------------------------------------


wt-amt1        TGATGCCGCTGTGGGGCTGAGGGTTAGCTCGCAGGAGGAGTACGTCGGTCTCGACCTGTC 1148
V156delta-rev  TGATGCCGCTGTGGGGCTGAGGGTTAGCTCGCAGGAGGAGTACGTCGGTCTCGACCTGTC 1369
V156delta-fwd  ------------------------------------------------------------


wt-amt1        GCAGCATGAGGAGGTTGCCTACACGTGA-------------------------------- 1176
V156delta-rev  GCAGCATGAGGAGGGTGCCTACACGCTCGAGCACCACCACCACCACCACTGAGATCCGGC 1429
V156delta-fwd  ------------------------------------------------------------


wt-amt1        ---------------------
V156delta-rev  TGTAACAAGCCGAAAGAGCGGTC 1452
V156delta-fwd  ---------------------
```

85

V156L

Scores table

```
SeqA Name        Len(nt)  SeqB Name         Len(nt)  Score
==========================================================
1    wt-amt1     1176     2    V156L-fwd     1055     94
1    wt-amt1     1176     3    V156L-rev     1466     97
2    V156L-fwd   1055     3    V156L-rev     1466     96
==========================================================
```

Alignment

```
wt-amt1     ------------------------------------------------------------
V156L-rev   CTTTTGGGAGCCCCCGGTGGGGATACCCGGGCCCACCAGAAAGGCTTTCCCGGGCTTTAG  60
V156L-fwd   ------------------------------------------------------------


wt-amt1     ------------------------------------------------------------
V156L-rev   AAGAGTATGGAGAATCTTCGGATTCCCCCGGGAAAATTAAATCCGGACTTCCATTTATAG  120
V156L-fwd   ------------------------------------------------------------


wt-amt1     ------------------------------------------------------------
V156L-rev   GGGGAAATTGGTGAACGGGAATAACCAATTTCCCCTTCTTTAGAAAATTAATTTTGTTTT  180
V156L-fwd   ------------------------------------------------GTGAGAATTCCCTCAG  16


wt-amt1     --------------------------------ATGAGTGACGGAAATGTCGC-ATGG-  24
V156L-rev   AAACTTTTAAGAAAGGGAGATAATACATATTGGAATGGACCGGAAAATGTTCGCCATGGG  240
V156L-fwd   AATATTTTGTTTACTTTAAGAAGGAGATATACATATGAGTGACGGAAATGTCCC-ATGG-  74


wt-amt1     ATACTCG-CAT--CCACGGCCCTT--GTAATG-CTGATGGTGCC--GGGAGTGGGGTTCT  76
V156L-rev   ATACTTGGCATTCCCACGGCCCCTTGGTAATGGTTGATGGTGCCCGGGAAGTGGGGTTCT  300
V156L-fwd   ATACTCG-CAT--CCACGGCCCTT--GTAATG-CTGATGGTGCC--GGGAGTGGGGTTCT  126


wt-amt1     TTT-ACGC--AGGAATGGTAAGG-AGAAAGAATGC-AGTTAACA-TGATTGCG-CTGAGC  129
V156L-rev   TTTTACGCCAGGAAATGGTAAGGGAGAAAGGATGCCAGTTAACAATGATTGCGGTTGAGC  360
V156L-fwd   TTT-ACGC--AGGAATGGTAAGG-AGAAAGAATGC-AGTTAACA-TGATTGCG-CTGAGC  179


wt-amt1     TT-CATATCA-CTCATAATCA-CGGTTTTGCTGTGGATATTCTACGGCTACT-CGGTGAG  185
V156L-rev   TTTCATATCAATTCATAATCAACGTTTTTGCTGTGGATATTCTACGGCTACTTCGGTGAG  420
V156L-fwd   TT-CATATCA-CTCATAATCA-CGGTTTTGCTGTGGATATTCTACGGCTACT-CGGTGAG  235


wt-amt1     CTTCGGAAATGACATCTCTGGAATCATTGGAGGGCTGAATTATGCACTGCTAAGCGGAGT  245
V156L-rev   CTTCGGGAATGACATCTCTGGAATCATTGGAGGGCTGAATTATGCACTGCTAAGCGGAGT  480
V156L-fwd   CTTCGGAAATGACATCTCTGGAATCATTGGAGGGCTGAATTATGCACTGCTAAGCGGAGT  295


wt-amt1     TAAGGGGGAGGATTTGCTGTTCATGATGTACCAGATGATGTTCGCCGCTGTCACAATTGC  305
V156L-rev   TAAGGGGGAGGATTTGCTGTTCATGATGTACCAGATGATGTTCGCCGCTGTCACAATTGC  540
V156L-fwd   TAAGGGGGAGGATTTGCTGTTCATGATGTACCAGATGATGTTCGCCGCTGTCACAATTGC  355


wt-amt1     AATCCTCACCTCCGCAATTGCTGAGAGAGCAAAAGTTTCATCGTTCATTCTCCTCAGCGC  365
V156L-rev   AATCCTCACCTCCGCAATTGCTGAGAGAGCAAAAGTTTCATCGTTCATTCTCCTCAGCGC  600
V156L-fwd   AATCCTCACCTCCGCAATTGCTGAGAGAGCAAAAGTTTCATCGTTCATTCTCCTCAGCGC  415


wt-amt1     TCTGTGGCTTACGTTCGTTTACGCCCCCTTCGCACACTGGCTTTGGGGTGGGGGGTGGCT  425
V156L-rev   TCTGTGGCTTACGTTCGTTTACGCCCCCTTCGCACACTGGCTTTGGGGTGGGGGGTGGCT  660
V156L-fwd   TCTGTGGCTTACGTTCGTTTACGCCCCCTTCGCACACTGGCTTTGGGGTGGGGGGTGGCT  475


wt-amt1     GGCAAAGCTCGGCGCCCTCGACTTTGCTGGAGGTATGGTTGTTCACATAAGCTCGGGATT  485
V156L-rev   GGCAAAGCTCGGCGCCCTCGACTTTGCTGGAGGTATGGTTCTTCACATAAGCTCGGGATT  720
V156L-fwd   GGCAAAGCTCGGCGCCCCTCGACTTTGCTGGAGGTATGGTTCTTCACATAAGCTCGGGATT  535
```

```
wt-amt1      TGCTGCACTTGCAGTCGCGATGACGATAGGTAAGAGGGCGGGATTCGAGGAGTACTCGAT 545
V156L-rev    TGCTGCACTTGCAGTCGCGATGACGATAGGTAAGAGGGCGGGATTCGAGGAGTACTCGAT 780
V156L-fwd    TGCTGCACTTGCAGTCGCGATGACGATAGGTAAGAGGGCGGGATTCGAGGAGTACTCGAT 595


wt-amt1      AGAGCCACACAGCATTCCGCTGACGCTCATTGGCGCTGCCCTGCTTTGGTTTGGGTGGTT 605
V156L-rev    AGAGCCACACAGCATTCCGCTGACGCTCATTGGCGCTGCCCTGCTTTGGTTTGGGTGGTT 840
V156L-fwd    AGAGCCACACAGCATTCCGCTGACGCTCATTGGCGCTGCCCTGCTTTGGTTTGGGTGGTT 655


wt-amt1      CGGATTCAACGGCGGAAGTGCATTGGCTGCAAACGATGTGGCCATCAACGCCGTGGTGGT 665
V156L-rev    CGGATTCAACGGCGGAAGTGCATTGGCTGCAAACGATGTGGCCATCAACGCCGTGGTGGT 900
V156L-fwd    CGGATTCAACGGCGGAAGTGCATTGGCTGCAAACGATGTGGCCATCAACGCCGTGGTGGT 715


wt-amt1      CACAAACACCTCAGCAGCAGTAGCAGGGTTTGTCTGGATGGTAATTGGATGGATTAAGGG 725
V156L-rev    CACAAACACCTCAGCAGCAGTAGCAGGGTTTGTCTGGATGGTAATTGGATGGATTAAGGG 960
V156L-fwd    CACAAACACCTCAGCAGCAGTAGCAGGGTTTGTCTGGATGGTAATTGGATGGATTAAGGG 775


wt-amt1      AAAGCCGGGGAGTCTTGGGATAGTGAGCGGTGCAATTGCTGGGCTTGCCGCCATAACCCC 785
V156L-rev    AAAGCCGGGGAGTCTTGGGATAGTGAGCGGTGCAATTGCTGGGCTTGCCGCCATAACCCC 1020
V156L-fwd    AAAGCCGGGGAGTCTTGGGATAGTGAGCGGTGCAATTGCTGGGCTTGCCGCCATAACCCC 835


wt-amt1      CGCAGCAGGCTTTGTGGATGTAAAGGGAGCGATTGTCATAGGTCTTGTGGCTGGAATAGT 845
V156L-rev    CGCAGCAGGCTTTGTGGATGTAAAGGGAGCGATTGTCATAGGTCTTGTGGCTGGAATAGT 1080
V156L-fwd    CGCAGCAGGCTTTGTGGATGTAAAGGGAGCGATTGTCATAGGTCTTGTGGCTGGAATAGT 895


wt-amt1      ATGCTACCTTGCTATGGACTTCAGAATAAAGAAGAAGATAGACGAGAGCCTTGATGCTTG 905
V156L-rev    ATGCTACCTTGCTATGGACTTCAGAATAAAGAAGAAGATAGACGAGAGCCTTGATGCTTG 1140
V156L-fwd    ATGCTACCTTGCTATGGACTTCAGAATAAAGAAGAAGATAGACGAGAGCCTTGATGCTTG 955


wt-amt1      GGCGATTCACGGAATAGGCGGTTTATGGGGAAGT-GTTGCAGTTGGC-ATTCTTGCAAAT 963
V156L-rev    GGCGATTCACGGAATAGGCGGTTTATGGGGAAGT-GTTGCAGTTGGC-ATTCTTGCAAAT 1198
V156L-fwd    GGCGATTCACGGAATAGGCGGTTTATGGGGGAATTGTTGCAGTTGGCCATCCTTGCAAAT 1015


wt-amt1      CCGGAGGTTAACGGATATGCAGGCCTACTGTTCGGAAATCCGCAACTGCTAGTTTCACAA 1023
V156L-rev    CCGGAGGTTAACGGATATGCAGGCCTACTGTTCGGAAATCCGCAACTGCTAGTTTCACAA 1258
V156L-fwd    CCAGAGGTTAACGGATATGAAGGCATACTGTTTGGAAAAA------------------- 1055


wt-amt1      CTGATTGCGGTTGCATCCACAACAGCCTACGCCTTCCTCGTGACGCTGATACTGGCAAAG 1083
V156L-rev    CTGATTGCGGTTGCATCCACAACAGCCTACGCCTTCCTCGTGACGCTGATACTGGCAAAG 1318
V156L-fwd    ------------------------------------------------------------


wt-amt1      GCTGTTGATGCCGCTGTGGGGCTGAGGGTTAGCTCGCAGGAGGAGTACGTCGGTCTCGAC 1143
V156L-rev    GCTGTTGATGCCGCTGTGGGGCTGAGGGTTAGCTCGCAGGAGGAGTACGTCGGTCTCGAC 1378
V156L-fwd    ------------------------------------------------------------


wt-amt1      CTGTCGCAGCATGAGGAGGTTGCCTACACGTGA--------------------------- 1176
V156L-rev    CTGTCGCAGCATGAGGAGGGGGCCTACACGCTCGAGCACCACCACCACCACCACTGAGAT 1438
V156L-fwd    ------------------------------------------------------------


wt-amt1      ---------------------------
V156L-rev    CCGGCTGTAACAAGCCGAAAGAGCGGCC 1466
V156L-fwd    ---------------------------
```

V156S

Scores table

```
SeqA Name          Len(nt)  SeqB Name          Len(nt)  Score
=============================================================
1    wt-amt1       1176     2    V156S-fwd     1149     93
1    wt-amt1       1176     3    V156S-rev     1054     79
2    V156S-fwd     1149     3    V156S-rev     1054     70
=============================================================
```

Alignment

```
wt-amt1      ------------------------------------------------------ATGAGTGA 8
V156S-fwd    GTGAGAATTCCCTCTAAAAAATTTTGTTTAACTTTAAGAAGGAGATATACATATGAGTGA 60
V156S-rev    ------------------------------------------------------------

wt-amt1      CGGAAATGTCGCATGGATACTCGCATCCACGGCCCTTGTAATGCTGATGGTGCCGGGAGT 68
V156S-fwd    CGGAAATGTCCCATGGATACTCGCATCCACGGCCCTTGTAATGCTGATGGTGCCGGGAGT 120
V156S-rev    ------------------------------------------------------------

wt-amt1      GGGGTTCTTTTACGCAGGAATGGTAAGGAGAAAGAATGCAGTTAACATGATTGCGCTGAG 128
V156S-fwd    GGGGTTCTTTTACGCAGGAATGGTAAGGAGAAAGAATGCAGTTAACATGATTGCGCTGAG 180
V156S-rev    ------------------------------------------------------------

wt-amt1      CTTCATATCACTCATAATCACGGTTTTGCTGTGGATATTCTACGGCTACTCGGTGAGCTT 188
V156S-fwd    CTTCATATCACTCATAATCACGGTTTTGCTGTGGATATTCTACGGCTACTCGGTGAGCTT 240
V156S-rev    ------------------------------------------------------------

wt-amt1      CGGAAATGACATCTCTGGAATCATTGGAGGGCTGAATTATGCACTGCTAAGCGGAGTTAA 248
V156S-fwd    CGGAAATGACATCTCTGGAATCATTGGAGGGCTGAATTATGCACTGCTAAGCGGAGTTAA 300
V156S-rev    --------------------ATGGGGGGTATGGAATTTTGTTTTTTTCAATGGATGT 38

wt-amt1      GGGGGAGGATTTGCTGTTCATGATGTACCAGATGATGTTCGCCGCTGTCACAATTGCAAT 308
V156S-fwd    GGGGGAGGATTTGCTGTTCATGATGTACCAGATGATGTTCGCCGCTGTCACAATTGCAAT 360
V156S-rev    TTCCCATGATTGATTGTTTATGC----CCGGCTG--GTCCACCAAATTGGCAAATCCCTT 92

wt-amt1      CCTCACCTCCGCAA------TTGCTGAGAGAGCAAAAGTTTCATCGTTCATTCT------ 356
V156S-fwd    CCTCACCTCCGCAA------TTGCTGAGAGAGCAAAAGTTTCATCGTTCATTCT------ 408
V156S-rev    CACCCTTCCTGCAAATTTGCCTGATGAGAAGCTAAAAAGTTTTCCATCCGTTTCATTTCT 152

wt-amt1      -CCTCAG--CGCTC-TGTGG--CTTACG--TTCGTTTACG---CCCCCTTCGC-ACACT- 403
V156S-fwd    -CCTCAG--CGCTC-TGTGG--CTTACG--TTCGTTTACG---CCCCCTTCGC-ACACT- 455
V156S-rev    CCCTCAAGCCGCTCCTGTAGGCTTTACGGTTCCGTTTACCGCCCCCCTTTCGCCACACTT 212

wt-amt1      GGCTTT---GGGGTGGGGGGTGGCT-GGCAAA-GCTCGGCGCCC-TCGACTTT-GCTGG- 455
V156S-fwd    GGCTTT---GGGGTGGGGGGTGGCT-GGCAAA-GCTCGGCGCCC-TCGACTTT-GCTGG- 507
V156S-rev    GGCTTTTGTGGGGTGGGGGGGTGGCTTGGCAAAAGCTCGGGGCCCCTCGACTTTTGCTGGG 272

wt-amt1      AGGTATGGTTGTTC-ACATAAGCTCGGGATTTGCTGCACTTGCAGTCGCGATGACGATAG 514
V156S-fwd    AGGTATGGTTAGCC-ACATAAGCTCGGGATTTGCTGCACTTGCAGTCGCGATGACGATAG 566
V156S-rev    AGGTATGGTTAGCCCACATAAGCTCGGGATTTGCTGCACTTGCAGTCGCGATGACGATAG 332

wt-amt1      GTAAGAGGGCGGGATTCGAGGAGTACTCGATAGAGCCACACAGCATTCCGCTGACGCTCA 574
V156S-fwd    GTAAGAGGGCGGGATTCGAGGAGTACTCGATAGAGCCACACAGCATTCCGCTGACGCTCA 626
V156S-rev    GTAAGAGGGCGGGATTCGAGGAGTACTCGATAGAGCCACACAGCATTCCGCTGACGCTCA 392

wt-amt1      TTGGCGCTGCCCTGCTTTGGTTTGGGTGGTTCGGATTCAACGGCGGAAGTGCATTGGCTG 634
```

```
V156S-fwd    TTGGCGCTGCCCTGCTTTGGTTTGGGTGGTTCGGATTCAACGGCGGAAGTGCATTGGCTG  686
V156S-rev    TTGGCGCTGCCCTGCTTTGGTTTGGGTGGTTCGGATTCAACGGCGGAAGTGCATTGGCTG  452


wt-amt1      CAAACGATGTGGCCATCAACGCCGTGGTGGTCACAAACACCTCAGCAGCAGTAGCAGGGT  694
V156S-fwd    CAAACGATGTGGCCATCAACGCCGTGGTGGTCACAAACACCTCAGCAGCAGTAGCAGGGT  746
V156S-rev    CAAACGATGTGGCCATCAACGCCGTGGTGGTCACAAACACCTCAGCAGCAGTAGCAGGGT  512


wt-amt1      TTGTCTGGATGGTAATTGGATGGATTAAGGGAAAGCCGGGGAGTCTTGGGATAGTGAGCG  754
V156S-fwd    TTGTCTGGATGGTAATTGGATGGATTAAGGGAAAGCCGGGGAGTCTTGGGATAGTGAGCG  806
V156S-rev    TTGTCTGGATGGTAATTGGATGGATTAAGGGAAAGCCGGGGAGTCTTGGGATAGTGAGCG  572


wt-amt1      GTGCAATTGCTGGGCTTGCCGCCATAACCCCCGCAGCAGGCTTTGTGGATGTAAAGGGAG  814
V156S-fwd    GTGCAATTGCTGGGCTTGCCGCCATAACCCCCGCAGCAGGCTTTGTGGATGTAAAGGGAG  866
V156S-rev    GTGCAATTGCTGGGCTTGCCGCCATAACCCCCGCAGCAGGCTTTGTGGATGTAAAGGGAG  632


wt-amt1      CGATTGTCATAGGTCTTGTGGCTGGAATAGTATGCTACCTTGCTATGGACTTCAGAATAA  874
V156S-fwd    CGATTGTCATAGGTCTTGTGGCTGGAATAGTATGCTACCTTGCTATGGACTTCAGAATAA  926
V156S-rev    CGATTGTCATAGGTCTTGTGGCTGGAATAGTATGCTACCTTGCTATGGACTTCAGAATAA  692


wt-amt1      AGAAGAAGATAGACGAGAGCCTTGATGCTTGGGCGATTCACGGAATAGGCGGTTTATGGG  934
V156S-fwd    AGAAGAAGATAGACGAGAGCCTTGATGCTTGGGCGATTCACGGAATAGGCGGTTTATGGG  986
V156S-rev    AGAAGAAGATAGACGAGAGCCTTGATGCTTGGGCGATTCACGGAATAGGCGGTTTATGGG  752


wt-amt1      GAAGTGTTGCAGTT-GGCATTCTTGCAAATCCGG-AGGTTAACGG-ATATGCA-GGCCTA  990
V156S-fwd    GAAGTGTTGCAGTTTGGCATTCTTGCAAATCCGGGAGGTTAACGGGATATGCAAGGCCTA 1046
V156S-rev    GAAGTGTTGCAGTT-GGCATTCTTGCAAATCCGG-AGGTTAACGG-ATATGCA-GGCCTA  808


wt-amt1      CTG-TTCGGAAA-TCCGCAA-CTGCTA-GTTTCACAACTG-ATTGCGG-TTGCATCCACA 1044
V156S-fwd    CTGGTTCGGAAAATCAGCAAACTGCTAAGTTTCACAACTGGATTGCGGGTTGCATCCACA 1106
V156S-rev    CTG-TTCGGAAA-TCCGCAA-CTGCTA-GTTTCACAACTG-ATTGCGG-TTGCATCCACA  862


wt-amt1      ACA-GCCTACGCCTTCCTC--GTGACGCTGATACT-GGCAAAGGCTGTTGATGCCGCTGT 1100
V156S-fwd    ACAAGCCTACGCCTTTCCTCGGGGACGCTGATACTTGGCCAAA---------------- 1149
V156S-rev    ACA-GCCTACGCCTTCCTC--GTGACGCTGATACT-GGCAAAGGCTGTTGATGCCGCTGT  918


wt-amt1      GGGGCTGAGGGTTAGCTCGCAGGAGGAGTACGTCGGTCTCGACCTGTCGCAGCATGAGGA 1160
V156S-fwd    ------------------------------------------------------------
V156S-rev    GGGGCTGAGGGTTAGCTCGCAGGAGGAGTACGTCGGTCTCGACCTGTCGCAGCATTTGGA  978


wt-amt1      GGTTGCCTACACGTGA-------------------------------------------- 1176
V156S-fwd    ------------------------------------------------------------
V156S-rev    GGGGGGGCCTACACGCTCGAGCACCACCACCACCACCACTGAGATCCGGCTGCTAACAAAG 1038


wt-amt1      ----------------
V156S-fwd    ----------------
V156S-rev    CCCGAAAGAAGGAGCN 1054
```

89

W201A

Scores table

```
SeqA Name          Len(nt)  SeqB Name          Len(nt)  Score
=============================================================
1    wt-amt1       1176     2    W201A-fwd     1110     92
1    wt-amt1       1176     3    W201A-rev     1303     68
2    W201A-fwd     1110     3    W201A-rev     1303     61
=============================================================
```

Alignment

```
wt-amt1     ------------------------------------------------------------ATG 3
W201A-fwd   NCCGTAAGAATTCCCTCTACAATATTTTGTTTAACTTTAAGAA--GGAGATATACATATG 58
W201A-rev   -----------TTCCCCCACCTTCCAAAATAAAAATTCCAACACGGGGGGGTTTTTTTTTG 49


wt-amt1     AGTGACGGAAATGTCGCATGGATACTCGCATCCAC-GGCCCTTGTAATGCTGATGGTGCC 62
W201A-fwd   AGTGACGGAAATGCCCCCTGAATACTCGCATCCAC-GGCCCTTGTAATGCTGATGGTGCC 117
W201A-rev   GGCCCATGGTTGGGGAATTAAATTTTTTCCTTAACCGGGGCTTTTAACTTTTCCGGGGTT 109


wt-amt1     GGGAGTGGGGTTCTTTTACGCAGGAATGGTAAGGAGAAAGAATGCAGTTAACATGATTGC 122
W201A-fwd   GGGAGTGGGGTTCTTTTACGCAGGAATGGTAAGGAGAAAGAATGCAGTTAACATGATTGC 177
W201A-rev   AGAAA---GGCATTTTCACGGGTAAAAATTTGGAACCAATCCTTCCTTGGGGAAAATTCC 166


wt-amt1     G---CTGAGCTTCATATCACTCATAATCACGGTTTTGCTGTGGATATTCTACGGCTACTC 179
W201A-fwd   G---CTGAGCTTCATATCACTCATAATCACGGTTTTGCTGTGGATATTCTACGGCTACTC 234
W201A-rev   AATTTTGGGAAGGGGGGCTTGGAAAATTTATTGGCCATCTTGCATTTAATGCGGTAAAGT 226


wt-amt1     GGTGAGCTTCGGAAATGACATCTCTGGAATCATTGGAGGGCTGAATTATGCACTGCTAAG 239
W201A-fwd   GGTGAGCTTCGGAAATGACATCTCTGGAATCATTGGAGGGCTGAATTATGCACTGCTAAG 294
W201A-rev   CTTATAGGGGGGGGGAAGGGATTTTTGGCCTAGTTTCCAA--TGAATGGTAACCCAAGAAT 284


wt-amt1     CGGAGTTAAGGGGGAGGATTTGCTGTTCATGATGTACCAGATGATGTTCGCCGCTGTCAC 299
W201A-fwd   CGGAGTTAAGGGGGAGGATTTGCTGTTCATGATGTACCAGATGATGTTCGCCGCTGTCAC 354
W201A-rev   TGAATGGTTTCGGCCCGGCTTGTTCCACA-AATTTGCCA-ATTCCCTCCAACCTTCCCGC 342


wt-amt1     AATTGCAATCCTC---ACCTCCGCAATTGCTGAGAGAGCAAAAGTTTCATCGTTCATTCT 356
W201A-fwd   AATTGCAATCCTC---ACCTCCGCAATTGCTGAGAGAGCAAAAGTTTCATCGTTCATTCT 411
W201A-rev   AAATTTGCTTGAAGAGAAGCAAAAAAGTTTTCCATCCGTTTCCATTTCTTCCTTCAAGCG 402


wt-amt1     CCTCA-GCGCTCTGTGGCTTACGTTCGTTTACGCCCCCTTCGC-ACAC-TGGCTTT--GG 411
W201A-fwd   CCTCA-GCGCTCTGTGGCTTACGTTCGTTTACGCCCCCTTCGC-ACAC-TGGCTTT--GG 466
W201A-rev   GCTCTTGTGGGCTTACGTTTCGTTTTACCGCCCCCCCCTTCGCCACACCTGGCTTTTGGG 462


wt-amt1     GGTGGGGGGT--GGCTGGCAAAG--CTCGGCGCCC-TCGA-CTTTGCTGGAGG-TATGGT 464
W201A-fwd   GGTGGGGGGT--GGCTGGCAAAG--CTCGGCGCCC-TCGA-CTTTGCTGGAGG-TATGGT 519
W201A-rev   GGTGGGGGGGTGGGCTGGCAAAAGCTTCGGCGCCCCTCGAGCTTTGCTGGAGGGTATGGT 522


wt-amt1     TGTTCACATAAGCTCGGGATTTGCTGCACTTGCAGTCGCGATGACGATAGGTAAGAGGGC 524
W201A-fwd   TGTTCACATAAGCTCGGGATTTGCTGCACTTGCAGTCGCGATGACGATAGGTAAGAGGGC 579
W201A-rev   TGTTCACATAAGCTCGGGATTTGCTGCACTTGCAGTCGCGATGACGATAGGTAAGAGGGC 582


wt-amt1     GGGATTCGAGGAGTACTCGATAGAGCCACACAGCATTCCGCTGACGCTCATTGGCGCTGC 584
W201A-fwd   GGGATTCGAGGAGTACTCGATAGAGCCACACAGCATTCCGCTGACGCTCATTGGCGCTGC 639
W201A-rev   GGGATTCGAGGAGTACTCGATAGAGCCACACAGCATTCCGCTGACGCTCATTGGCGCTGC 642


wt-amt1     CCTGCTTTGGTTTGGGTGGTTCGGATTCAACGGCGGAAGTGCATTGGCTGCAAACGATGT 644
```

```
W201A-fwd    CCTGCTTTGGTTTGGGGCGTTCGGATTCAACGGCGGAAGTGCATTGGCTGCAAACGATGT 699
W201A-rev    CCTGCTTTGGTTTGGGGCGTTCGGATTCAACGGCGGAAGTGCATTGGCTGCAAACGATGT 702


wt-amt1      GGCCATCAACGCCGTGGTGGTCACAAACACCTCAGCAGCAGTAGCAGGGTTTGTCTGGAT 704
W201A-fwd    GGCCATCAACGCCGTGGTGGTCACAAACACCTCAGCAGCAGTAGCAGGGTTTGTCTGGAT 759
W201A-rev    GGCCATCAACGCCGTGGTGGTCACAAACACCTCAGCAGCAGTAGCAGGGTTTGTCTGGAT 762


wt-amt1      GGTAATTGGATGGATTAAGGGAAAGCCGGGGAGTCTTGGGATAGTGAGCGGTGCAATTGC 764
W201A-fwd    GGTAATTGGATGGATTAAGGGAAAGCCGGGGAGTCTTGGGATAGTGAGCGGTGCAATTGC 819
W201A-rev    GGTAATTGGATGGATTAAGGGAAAGCCGGGGAGTCTTGGGATAGTGAGCGGTGCAATTGC 822


wt-amt1      TGGGCTTGCCGCCATAACCCCCGCAGCAGGCTTTGTGGATGTAAAGGGAGCGATTGTCAT 824
W201A-fwd    TGGGCTTGCCGCCATAACCACCGCAGCAGGCTTTGTGGATGTAAAGGGAGCGATTGTCAT 879
W201A-rev    TGGGCTTGCCGCCATAACCCCCGCAGCAGGCTTTGTGGATGTAAAGGGAGCGATTGTCAT 882


wt-amt1      AGGTCTTGTGGCTGGAATAGTATGCTACCTTGCTATGGACTTCAGAATAAAGAAGAAGAT 884
W201A-fwd    AGGTCTTGTGGCTGGAATAGTATGCTACCTTGCTATGGACTTCAGAATAAAGAAGAAGAT 939
W201A-rev    AGGTCTTGTGGCTGGAATAGTATGCTACCTTGCTATGGACTTCAGAATAAAGAAGAAGAT 942


wt-amt1      AGACGAGAGCCTTGATGCTTGGGCGATTCACGGAATAGGCGGTTTATGGGGAA-GTGTTG 943
W201A-fwd    AGACGAGAGCCTTGATGCTTGGGCGATTCACGGAATAGGCGGTTTATGGGGAAAGTGTTG 999
W201A-rev    AGACGAGAGCCTTGATGCTTGGGCGATTCACGGAATAGGCGGTTTATGGGGAA-GTGTTG 1001


wt-amt1      CAGTTGGCATTCTTG-CAAATCCGG-AGGTTAACGG-ATATGCA-GGCCTACTG-TTCGG 998
W201A-fwd    CAGTTGGCATTCCTGGCAAATCCGGGAGGTTAACGGGATATGCAAGGCCTACTGGTTCGG 1059
W201A-rev    CAGTTGGCATTCTTG-CAAATCCGG-AGGTTAACGG-ATATGCA-GGCCTACTG-TTCGG 1056


wt-amt1      AAATCC-GCAACT-GCTAGTTT-CACAACT-GATTGCGG-TTGCATCCACAACAGCCTAC 1053
W201A-fwd    AAATCCCGCAACTTGCTAGTTTTCACAACTTGATTGCGGGTTGCCTTCCCC--------- 1110
W201A-rev    AAATCC-GCAACT-GCTAGTTT-CACAACT-GATTGCGG-TTGCATCCACAACAGCCTAC 1111


wt-amt1      GCCTTCCTCGTGACGCTGATACTGGCAAAGGCTGTTGATGCCGCTGTGGGGCTGAGGGTT 1113
W201A-fwd    ------------------------------------------------------------
W201A-rev    GCCTTCCTCGTGACGCTGATACTGGCAAAGGCTGTTGATGCCGCTGTGGGGCTGAGGGTT 1171


wt-amt1      AGCT-CGCAGGAGGAGTA-CGTCGGTCTCGACCTGTCGCAGCATGAG-GAGGTTGCCTAC 1170
W201A-fwd    ------------------------------------------------------------
W201A-rev    AGCTGCGCAGGACGACTACCGTCGGTCTCGACCTGTCGCAGCCTTTTTAAGGGGGCCTAC 1231


wt-amt1      ACGTGA------------------------------------------------------ 1176
W201A-fwd    ------------------------------------------------------------
W201A-rev    ACGCTCGAGCACACACCACGCACCGCTCACTGAGATCCGGCTACTAACAAAGCCCGAAAG 1291


wt-amt1      ------------
W201A-fwd    ------------
W201A-rev    AAGGGGTTCGTN 1303
```

W201G

Scores table

```
SeqA Name          Len(nt)  SeqB Name          Len(nt)   Score
=============================================================
1    wt-amt1       1176     2    W201G-fwd     1082      92
1    wt-amt1       1176     3    W201G-rev     1371      98
2    W201G-fwd     1082     3    W201G-rev     1371      97
=============================================================
```

Alignment

```
wt-amt1     ------------------------------------------------------------
W201G-rev   CCGGGGTTAAAGGGATCGAGATACTTCGTTTCCCCGGAAAATTAATTCGGATTCACTTAT  60
W201G-fwd   ------------------------------------------------------------


wt-amt1     ------------------------------------------------------------
W201G-rev   AGGGGGATTTGTGAGCGGATAACCAATTCCCTTTCTGGAATTAATTTTGTTTAACTTTAA 120
W201G-fwd   -------------CGTTTGAAAGGGTCCATTTCCCTCTAGAATAATTTTGTTTAACTTTA   47


wt-amt1     ----------------ATG-AGTGACGGAAATGTCG-CATGGATACTCGCAT-CCACGGC  41
W201G-rev   GAAAGGAGATTTCCATATGGAGTACCGGAAATGTTGGCATGGATATTCGCATTCCAGGGC 180
W201G-fwd   AGAAGGAGATATACATATG-AGTGACGGAAATGTTG--ATGGGTACTCGCAT-CCACGGC 103


wt-amt1     CCTTGTAATGCTGATGGTGCC-GGGAGTGGGGTTCTTTTACGCAGGAATGGTAAGGAGAA 100
W201G-rev   CCTTGTAATGCTGATGGTGCCCGGGAGTGGGGTTCTTTTACGCAGGAATGGTAAGGAGAA 240
W201G-fwd   CCTTGTAATGCTGATGGTGCC-GGGAGGGGGGTTCTTTTACGCAGGAATGGTAAGGAGAA 162


wt-amt1     AGAATGCAGTTAACATGATTGCGCTGAGCTTCATATCACTCATAATCACGGTTTTGCTGT 160
W201G-rev   AGAATGCAGTTAACATGATTGCGCTGAGCTTCATATCACTCATAATCACGGTTTTGCTGT 300
W201G-fwd   AGAATGCAGTTAACATGATTGCGCTGAGCTTCATATCACTCATAATCACGGTTTTGCTGT 222


wt-amt1     GGATATTCTACGGCTACTCGGTGAGCTTCGGAAATGACATCTCTGGAATCATTGGAGGGC 220
W201G-rev   GGATATTCTACGGCTACTCGGTGAGCTTCGGAAATGACATCTCTGGAATCATTGGAGGGC 360
W201G-fwd   GGATATTCTACGGCTACTCGGTGAGCTTCGGAAATGACATCTCTGGAATCATTGGAGGGC 282


wt-amt1     TGAATTATGCACTGCTAAGCGGAGTTAAGGGGGAGGATTTGCTGTTCATGATGTACCAGA 280
W201G-rev   TGAATTATGCACTGCTAAGCGGAGTTAAGGGGGAGGATTTGCTGTTCATGATGTACCAGA 420
W201G-fwd   TGAATTATGCACTGCTAAGCGGAGTTAAGGGGGAGGATTTGCTGTTCATGATGTACCAGA 342


wt-amt1     TGATGTTCGCCGCTGTCACAATTGCAATCCTCACCTCCGCAATTGCTGAGAGAGCAAAAG 340
W201G-rev   TGATGTTCGCCGCTGTCACAATTGCAATCCTCACCTCCGCAATTGCTGAGAGAGCAAAAG 480
W201G-fwd   TGATGTTCGCCGCTGTCACAATTGCAATCCTCACCTCCGCAATTGCTGAGAGAGCAAAAG 402


wt-amt1     TTTCATCGTTCATTCTCCTCAGCGCTCTGTGGCTTACGTTCGTTTACGCCCCCTTCGCAC 400
W201G-rev   TTTCATCGTTCATTCTCCTCAGCGCTCTGTGGCTTACGTTCGTTTACGCCCCCTTCGCAC 540
W201G-fwd   TTTCATCGTTCATTCTCCTCAGCGCTCTGTGGCTTACGTTCGTTTACGCCCCCTTCGCAC 462


wt-amt1     ACTGGCTTTGGGGTGGGGGGTGGCTGGCAAAGCTCGGCGCCCTCGACTTTGCTGGAGGTA 460
W201G-rev   ACTGGCTTTGGGGTGGGGGGTGGCTGGCAAAGCTCGGCGCCCTCGACTTTGCTGGAGGTA 600
W201G-fwd   ACTGGCTTTGGGGTGGGGGGTGGCTGGCAAAGCTCGTCGCCCTCGACTTTGCTGGAGGTA 522


wt-amt1     TGGTTGTTCACATAAGCTCGGGATTGCTGCACTTGCAGTCGCGATGACGATAGGTAAGA 520
W201G-rev   TGGTTGTTCACATAAGCTCGGGATTGCTGCACTTGCAGTCGCGATGACGATAGGTAAGA 660
W201G-fwd   TGGTTGTTCACATAAGCTCGGGATTGCTGCACTTGCAGTCGCGATGACGATAGGTAAGA 582


wt-amt1     GGGCGGGATTCGAGGAGTACTCGATAGAGCCACACAGCATTCCGCTGACGCTCATTGGCG 580
W201G-rev   GGGCGGGATTCGAGGAGTACTCGATAGAGCCACACAGCATTCCGCTGACGCTCATTGGCG 720
W201G-fwd   GGGCGGGATTCGAGGAGTACTCGATAGAGCCACACAGCATTCCGCTGACGCTCATTGGCG 642
```

```
wt-amt1    CTGCCCTGCTTTGGTTTGGGTGGTTCGGATTCAACGGCGGAAGTGCATTGGCTGCAAACG 640
W201G-rev  CTGCCCTGCTTTGGTTTGGGGGGTTCGGATTCAACGGCGGAAGTGCATTGGCTGCAAACG 780
W201G-fwd  CTGCCCTGCTTTGGTTTGGGGGGTTCGGATTCAACGGCGGAAGTGCATTGGCTGCAAACG 702


wt-amt1    ATGTGGCCATCAACGCCGTGGTGGTCACAAACACCTCAGCAGCAGTAGCAGGG-TTTGTC 699
W201G-rev  ATGTGGCCATCAACGCCGTGGTGGTCACAAACACCTCAGCAGCAGTAGCAGGG-TTTGTC 839
W201G-fwd  ATGTGGCCATCAACGCCGTGGTGGTCACAAACACCTCAGCAGCAGTAGCAGGGGTTTGTC 762


wt-amt1    TGGATGGTAATTGGATGGATTAAGGGAAAGCCGGGG-AGTCTTGGGATAGTGAGCGGTGC 758
W201G-rev  TGGATGGTAATTGGATGGATTAAGGGAAAGCCGGGG-AGTCTTGGGATAGTGAGCGGTGC 898
W201G-fwd  TGGATGGTAATTGGATGGATTAAGGGAAAGCCGGGGGAGTCTTGGGATAGTGAGCGGTGC 822


wt-amt1    AATTGCTGGGCTTGCCGCCATAACCCCCGCAGCAGGCTTTGTGGATGTAAAGGGAGCGAT 818
W201G-rev  AATTGCTGGGCTTGCCGCCATAACCCCCGCAGCAGGCTTTGTGGATGTAAAGGGAGCGAT 958
W201G-fwd  AATTGCTGGGCTTGCCGCCATAACCCCCGCAGCAGGCTTTGTGGATGTAAAGGGAGCGAT 882


wt-amt1    TGTCATAGGTCTTGTGGCTGGAATAGTATGCTACCTTGCTATGGACTTCAGAATAAAGAA 878
W201G-rev  TGTCATAGGTCTTGTGGCTGGAATAGTATGCTACCTTGCTATGGACTTCAGAATAAAGAA 1018
W201G-fwd  TGTCATAGGTCTTGTGGCTGGAATAGTATGCTACCTTGCTATGGACTTCAGAATAAAGAA 942


wt-amt1    GAAGATAGACGAGAGCCTTGATGCTTGGGCGATTCACGGAATAGGCGGTTTATGGGGAAG 938
W201G-rev  GAAGATAGACGAGAGCCTTGATGCTTGGGCGATTCACGGAATAGGCGGTTTATGGGGAAG 1078
W201G-fwd  GAAGATAGACGAGAGCCTTGATGCTTGGGCGATTCACGGAATAGGCGGTTTATGGGGAAG 1002


wt-amt1    TGTTGCAGTTGGCATTCTTGCAAATCCGGAGGTTAACGGATA-TGCAGGCCTACT-GTTC 996
W201G-rev  TGTTGCAGTTGGCATTCTTGCAAATCCGGAGGTTAACGGATA-TGCAGGCCTACT-GTTC 1136
W201G-fwd  TGTTGCAGTTGGCATTCTTGCAA-TCCGGAGGTTAACGGATAATGCAGGCCTACTTGTTC 1061


wt-amt1    GGAAATCCGCAACTGCTAGTTTCACAACTGATTGCGGTTGCATCCACAACAGCCTACGCC 1056
W201G-rev  GGAAATCCGCAACTGCTAGTTTCACAACTGATTGCGGTTGCATCCACAACAGCCTACGCC 1196
W201G-fwd  GGAAATCCGCAACTGCTTAGT--------------------------------------- 1082


wt-amt1    TTCCTCGTGACGCTGATACTGGCAAAGGCTGTTGATGCCGCTGTGGGGCTGAGGGTTAGC 1116
W201G-rev  TTCCTCGTGACGCTGATACTGGCAAAGGCTGTTGATGCCGCTGTGGGGCTGAGGGTTAGC 1256
W201G-fwd  ------------------------------------------------------------


wt-amt1    TCGCAGGAGGAGTACGTCGGTCTCGACCTGTCGCAGCATGAGGAGGTTGCCTACACGTGA 1176
W201G-rev  TCGCAGGAGGAGTACGTCGGTCTCGACCTGTCGCAGCATGAGGAGGGGGCCTACACGCTC 1316
W201G-fwd  ------------------------------------------------------------


wt-amt1    ------------------------------------------------------------
W201G-rev  GAGCACCACCACCACCACCACTGAGATCCGGCTGTAACAAGCCGAAAGAGCGGCC 1371
W201G-fwd  ------------------------------------------------------------
```

93

D149A

Scores table

```
SeqA Name        Len(nt)  SeqB Name        Len(nt)  Score
=========================================================
1    wt-amt1     1176     2    D149A-fwd   1035     93
1    wt-amt1     1176     3    D149A-rev   1147     90
2    D149A-fwd   1035     3    D149A-rev   1147     82
=========================================================
```

Alignment

```
wt-amt1     ------------------------------------------------ATGAGTGACGGAAAT 15
D149A-fwd   TAGAATTCCTCTATATATTTAGGCAGCTTTAATATGAGATTACATCTGACTGACTGAAAT 60
D149A-rev   ------------------------------------------------------------


wt-amt1     GTCGCATGGATACTCGCATCCACGGCCCTTGTAATGCTGATGGTGCCGGGAGTGGGGTTC 75
D149A-fwd   GTCCCCTTAAAAAGCCCATCCAAGGCCCTTGTAATGCTGATGGAGCCGGGAGTGGGGTTC 120
D149A-rev   ------------------------------------------------------------


wt-amt1     TTTTACGCAGGAATGGTAAGGAGAAAGAATGCAGTTAACATGATTGCGCTGAGCTTCATA 135
D149A-fwd   TTTTACGCACGAATGGTAAGGAGAAAGAATGCAGTTAACATGATTGCGCTGAGCTTCATA 180
D149A-rev   ----------------------------------TGGCGGCGGAGCATTCAATATCAAA 25


wt-amt1     TCACTCATAATCACGGTTTTGCT-GTGGA-TATTCT-ACGGC-TACT-CGGTGA-GCTTC 189
D149A-fwd   TCACTCATAATCACGGTTTTGCT-GTGGA-TATTCT-ACGGC-TACT-CGGTGA-GCTTC 234
D149A-rev   TCCATAATCCACGGGTTTTTGCTTGTGGAATATTCTTACGGGATACTTCGGTGAAGCTTC 85


wt-amt1     GG-AAATGACATCT-CTGGAATC--ATTGGAGGGCTGAATTA-TGCACT-GCTAAGCGG- 242
D149A-fwd   TT-AAATGACATCT-CTGGAATC--TTTGGAGGGCTGAATTA-TGCACT-GCTAAGCGG- 287
D149A-rev   GGGAAATGACATTTTCTGGAATCCATTGGAGGGGCTGAAATAATGCACTTGCTAAGCGGG 145


wt-amt1     AGTTAAGGGGGAGGA--TTTGCTGTTC-ATGATGTACCAGA-TGATGTTCGCCGCTGTCA 298
D149A-fwd   AGTTAAGGGGGAGGA--TTTGCTGTTC-ATGATGTACCAGA-TGATGTTCGCCGCTGTCA 343
D149A-rev   AGTTAAGGGGGAGGATTTTGCTGTTCCATGATGTACCAGAATGATGTTCGCCGCTGTCA 205


wt-amt1     CAA-TTGCAATCCTCACCTCCGCAATTGCTGAGAGAGCAAAAGTTTCATCGTTCATTCTC 357
D149A-fwd   CAA-TTGCAATCCTCACCTCCGCAATTGCTGAGAGAGCAAAAGTTTCATCGTTCATTCTC 402
D149A-rev   CAAATTGCAATCCTCACCTCCGCAATTGCTGAGAGAGCAAAAGTTTCATCGTTCATTCTC 265


wt-amt1     CTCAGCGCTCTGTGGCTTACGTTCGTTTACGCCCCCTTCGCACACTGGCTTTGGGGTGGG 417
D149A-fwd   CTCAGCGCTCTGTGGCTTACGTTCGTTTACGCCCCCTTCGCACACTGGCTTTGGGGTGGG 462
D149A-rev   CTCAGCGCTCTGTGGCTTACGTTCGTTTACGCCCCCTTCGCACACTGGCTTTGGGGTGGG 325


wt-amt1     GGGTGGCTGGCAAAGCTCGGCGCCCTCGACTTTGCTGGAGGTATGGTTGTTCACATAAGC 477
D149A-fwd   GGGTGGCTGGCAAAGCTCGGCGCCCTCGCCTTTGCTGGAGGTATGGTTGTTCACATAAGC 522
D149A-rev   GGGTGGCTGGCAAAGCTCGGCGCCCTCGCCTTTGCTGGAGGTATGGTTGTTCACATAAGC 385


wt-amt1     TCGGGATTTGCTGCACTTGCAGTCGCGATGACGATAGGTAAGAGGGCGGGATTCGAGGAG 537
D149A-fwd   TCGGGATTTGCTGCACTTGCAGTCGCGATGACGATAGGTAAGAGGGCGGGATTCCAGGAG 582
D149A-rev   TCGGGATTTGCTGCACTTGCAGTCGCGATGACGATAGGTAAGAGGGCGGGATTCGAGGAG 445


wt-amt1     TACTCGATAGAGCCACACAGCATTCCGCTGACGCTCATTGGCGCTGCCCTGCTTTGGTTT 597
D149A-fwd   TACTCGATAGAGCCACACAGCATTCCGCTGACGCTCATTGGCGCTGCCCTGCTTTGGTTT 642
D149A-rev   TACTCGATAGAGCCACACAGCATTCCGCTGACGCTCATTGGCGCTGCCCTGCTTTGGTTT 505


wt-amt1     GGGTGGTTCGGATTCAACGGCGGAAGTGCATTGGCTGCAAACGATGTGGCCATCAACGCC 657
D149A-fwd   GGGTGGTTCGGATTCAACGGCGGAAGTGCATTGGCTGCAAACGATGTGGCCATCAACGCC 702
D149A-rev   GGGTGGTTCGGATTCAACGGCGGAAGTGCATTGGCTGCAAACGATGTGGCCATCAACGCC 565
```

```
wt-amt1     GTGGTGGTCACAAACACCTCAGCAGCAGTAGCAGGGTTTGTCTGGATGGTAATTGGATGG 717
D149A-fwd   GTGGTGGTCACAAACACCTCAGCAGCAGTAGCAGGGTTTGTCTGGATGGTAATTGGATGG 762
D149A-rev   GTGGTGGTCACAAACACCTCAGCAGCAGTAGCAGGGTTTGTCTGGATGGTAATTGGATGG 625


wt-amt1     ATTAAGGGAAAGCCGGGGAGTCTTGGGATAGTGAGCGGTGCAATTGCTGGGCTTGCCGCC 777
D149A-fwd   ATTAAGGGAAAGCCGGCGAGTCTTGGGATAGTGAGCGGTGCAATTGCTGGGCTTGCCGCC 822
D149A-rev   ATTAAGGGAAAGCCGGGGAGTCTTGGGATAGTGAGCGGTGCAATTGCTGGGCTTGCCGCC 685


wt-amt1     ATAACCCCCGCAGCAGGCTTTGTGGATGTAAAGGGAGCGATTGTCATAGGTCTTGTGGCT 837
D149A-fwd   ATAACCCCCGCAGCAGGCTTTGTGGATGTAAAGGGAGCGATTGTCATAGGTCTTGTGGCT 882
D149A-rev   ATAACCCCCGCAGCAGGCTTTGTGGATGTAAAGGGAGCGATTGTCATAGGTCTTGTGGCT 745


wt-amt1     GGAATAGTATGCTACCTTGCTATGGACTTCAGAATAAAGAAGAAGATAGACGAGAGCCTT 897
D149A-fwd   GGAATAGTATGCTACCTTGCTATGGACTTCAGAATAAAGAAGAAGATAGACGAGAGCCTT 942
D149A-rev   GGAATAGTATGCTACCTTGCTATGGACTTCAGAATAAAGAAGAAGATAGACGAGAGCCTT 805


wt-amt1     GATGCTTGGGCGATTCACGGAATAGGCGGTTTATGGGGAAGTGTTGCAGTTGGCATTCTT 957
D149A-fwd   GATGCTTGGGCGATTCACGGAATAGGCGGTTTATGGGGAAGTGTTGCAGTTGGCATTCTT 1002
D149A-rev   GATGCTTGGGCGATTCACGGAATAGGCGGTTTATGGGGAAGTGTTGCAGTTGGCATTCTT 865


wt-amt1     GCAAATCCGGAGGTTAACGGATATGCAGGCCTACTGTTCGGAAATCCGCAACTGCTAGTT 1017
D149A-fwd   GCAAATCCGGAGGTTAACGGATATGCAGGCCTA--------------------------- 1035
D149A-rev   GCAAATCCGGAGGTTAACGGATATGCAGGCCTACTGTTCGGAAATCCGCAACTGCTAGTT 925


wt-amt1     TCACAACTGATTGCGGTTGCATCCACAACAGCCTACGCCTTCCTCGTGACGCTGATACTG 1077
D149A-fwd   ------------------------------------------------------------
D149A-rev   TCACAACTGATTGCGGTTGCATCCACAACAGCCTACGCCTTCCTCGTGACGCTGATACTG 985


wt-amt1     GCAAAGGCTGTTGATGCCGCTGTGGGGCTGAGGGTTAGCT-CGCAGGAGGAGTACGTCGG 1136
D149A-fwd   ------------------------------------------------------------
D149A-rev   GCAAAGGCTGTTGATGCCGCTGTGGGGCTGAGGGTTAGAGGCGCAGGAGGAGTACGTCGG 1045


wt-amt1     TCTCGACCTGTCGCAGCATGAGGAGGTTGCCTACACGTGA-------------------- 1176
D149A-fwd   ------------------------------------------------------------
D149A-rev   TCTCGACCTGTCGCACCTTTTTAGGGGGGGCCTACACGCTCGAGCACCACTCACCACCAC 1105


wt-amt1     --------------------------------------------
D149A-fwd   --------------------------------------------
D149A-rev   CACTGAGATCCGGCTGCTAACAAACCCGAAGGAGGGTGGAGN 1147
```

G309S

Scores table

```
SeqA Name          Len(nt)  SeqB Name          Len(nt)  Score
==============================================================
1     wt-amt1      1176      2    G309S-fwd     896      90
1     wt-amt1      1176      3    G309S-rev     1377     98
2     G309S-fwd    896       3    G309S-rev     1377     95
==============================================================
```

Alignment

```
wt-amt1    ------------------------------------------------------------
G309S-rev  AAACGTTTCCCGCGTTAAAGGAATCGGAAATCTTGATCCCCGGGAAATTAAATCGGACTC 60
G309S-fwd  ------------------------------------------------------------


wt-amt1    ------------------------------------------------------------
G309S-rev  CCTATTAGGGAAATTGGACGGGGTAACCAATTCCCTTCTGAGAAATATTTTGGTTTAACT 120
G309S-fwd  -------------------------GTGAGAAATTCCCTCTAGAAAATTTTGTTTAAC 33


wt-amt1    -------------------ATGAGTGACGG-AAATGTCGCATGGATACT-CGCATCCAC 38
G309S-rev  TTAAGAAGGGAGATTAACATTTGAAGGACGGGAAAGGTCGCATGGATACTTCGCTTCCAC 180
G309S-fwd  TTTAAGAAGGAGATATACATATGAGTGACGG-ACCCGTCACATGGATACT-CGCATCCAC 91


wt-amt1    GG-CCCTTGTAATGCTGATGGTGCCGGGAGTGGGGTTCTTTTACGCAGGAATGGTAAGGA 97
G309S-rev  GGGCCTTTGTAATGCTTATGGTGCCCGGAGTTGGGTTCTTTTACGCAGGAATGGTAAGGA 240
G309S-fwd  GG-CCCTTGTAATGCTGATGGTGCCGGGAGTGGGGTTCTTTTACGCAGGAATGGTAAGGA 150


wt-amt1    GAAAGAATGCAGTTAACATGATTGCGCTGAGCTTCATATCACTCATAATCACGGTTTTGC 157
G309S-rev  GAAAGAATGCAGTTAACATGATTGCGCTGAGCTTCATATCACTCATAATCACGGTTTTGC 300
G309S-fwd  GAAAGAATGCAGTTAACATGATTGCGCTGAGCTTCATATCACTCATAATCACGGTTTTGC 210


wt-amt1    TGTGGATATTCTACGGCTACTCGGTGAGCTTCGGAAATGACATCTCTGGAATCATTGGAG 217
G309S-rev  TGTGGATATTCTACGGCTACTCGGTGAGCTTCGGAAATGACATCTCTGGAATCATTGGAG 360
G309S-fwd  TGTGGATATTCTACGGCTACTCGGTGAGCTTCGGAAATGACATCTCTGGAATCATTGGAG 270


wt-amt1    GGCTGAATTATGCACTGCTAAGCGGAGTTAAGGGGGAGGATTTGCTGTTCATGATGTACC 277
G309S-rev  GGCTGAATTATGCACTGCTAAGCGGAGTTAAGGGGGAGGATTTGCTGTTCATGATGTACC 420
G309S-fwd  GGCTGAATTATGCACTGCTAAGCGGAGTTAAGGGGGAGGATTTGCTGTTCATGATGTACC 330


wt-amt1    AGATGATGTTCGCCGCTGTCACAATTGCAATCCTCACCTCCGCAATTGCTGAGAGAGCAA 337
G309S-rev  AGATGATGTTCGCCGCTGTCACAATTGCAATCCTCACCTCCGCAATTGCTGAGAGAGCAA 480
G309S-fwd  AGATGATGTTCGCCGCTGTCACAATTGCAATCCTCACCTCCGCAATTGCTGAGAGAGCAA 390


wt-amt1    AAGTTTCATCGTTCATTCTCCTCAGCGCTCTGTGGCTTACGTTCGTTTACGCCCCCTTCG 397
G309S-rev  AAGTTTCATCGTTCATTCTCCTCAGCGCTCTGTGGCTTACGTTCGTTTACGCCCCCTTCG 540
G309S-fwd  AAGTTTCATCGTTCATTCTCCTCAGCGCTCTGTGGCTTACGTTCGTTTACGCCCCCTTCG 450


wt-amt1    CACACTGGCTTTGGGGTGGGGGGTGGCTGGCAAAGCTCGGCGCCCTCGACTTTGCTGGAG 457
G309S-rev  CACACTGGCTTTGGGGTGGGGGGTGGCTGGCAAAGCTCGGCGCCCTCGACTTTGCTGGAG 600
G309S-fwd  CACACTGGCTTTGGGGTGGGGGGTGGCTGGCAAAGCTCGGCGCCCTCGACTTTGCTGGAG 510


wt-amt1    GTATGGTTGTTCACATAAGCTCGGGATTTGCTGCACTTGCAGTCGCGATGACGATAGGTA 517
G309S-rev  GTATGGTTGTTCACATAAGCTCGGGATTTGCTGCACTTGCAGTCGCGATGACGATAGGTA 660
G309S-fwd  GTATGGTTGTTCACATAAGCTCGGGATTTGCTGCACTTGCAGTCGCGATGACGATAGGTA 570


wt-amt1    AGAGGGCGGGATTCGAGGAGTACTCGATAGAGCCACACAGCATTCCGCTGACGCTCATTG 577
G309S-rev  AGAGGGCGGGATTCGAGGAGTACTCGATAGAGCCACACAGCATTCCGCTGACGCTCATTG 720
G309S-fwd  AGAGGGCGGGATTCGAGGAGTACTCGATAGAGCCACACAGCATTCCGCTGACGCTCATTG 630
```

```
wt-amt1     GCGCTGCCCTGCTTTGGTTTGGGTGGTTCGGATTCAACGGCGGAAGTGCATTGGCTGCAA 637
G309S-rev   GCGCTGCCCTGCTTTGGTTTGGGTGGTTCGGATTCAACGGCGGAAGTGCATTGGCTGCAA 780
G309S-fwd   GCGCTGCCCTGCTTTGGTTTGGGTGGTTCGGATTCAACGGCGGAAGTGCATTGGCTGCAA 690


wt-amt1     ACGATGTGGCCATCAACGCCGTGGTGGTCACAAACACCTCAGCAGCAGTAGCAGGGTTTG 697
G309S-rev   ACGATGTGGCCATCAACGCCGTGGTGGTCACAAACACCTCAGCAGCAGTAGCAGGGTTTG 840
G309S-fwd   ACGATGTGGCCATCAACGCCGTGGTGGTCACAAACACCTCAGCAGCAGTAGCAGGGTTTG 750


wt-amt1     TCTGGATGGTAATTGGATGGATTAAGGGAAAGCCGGG-GAGTCTT-GGGATAGTGA-GCG 754
G309S-rev   TCTGGATGGTAATTGGATGGATTAAGGGAAAGCCGGG-GAGTCTT-GGGATAGTGA-GCG 897
G309S-fwd   TCTGGATGGTAATTGGATGGATTAAGGGAAAGTCGAGTGAGTCTTTGGGATAGTGAAGAG 810


wt-amt1     GTGCAATTGC-TGGGCTTGCCGCCAT-AACCCCCGCAGC-AGGCTTT-GTGGATGTAAAG 810
G309S-rev   GTGCAATTGC-TGGGCTTGCCGCCAT-AACCCCCGCAGC-AGGCTTT-GTGGATGTAAAG 953
G309S-fwd   GTGAAATTGCATGAGCTTGTCGCCATTAACCCCCACAACTATGCTTTAGTGGATGAAAAG 870


wt-amt1     GG-AGCGATTGTCATAGGTCTTGTGGCTGGAATAGTATGCTACCTTGCTATGGACTTCAG 869
G309S-rev   GG-AGCGATTGTCATAGGTCTTGTGGCTGGAATAGTATGCTACCTTGCTATGGACTTCAG 1012
G309S-fwd   GGGAGGGATTGTCCTCCGTTCTTTTG---------------------------------- 896


wt-amt1     AATAAAGAAGAAGATAGACGAGAGCCTTGATGCTTGGGCGATTCACGGAATAGGCGGTTT 929
G309S-rev   AATAAAGAAGAAGATAGACGAGAGCCTTGATGCTTGGGCGATTCACGGAATAGGC[AGT]TT 1072
G309S-fwd   ------------------------------------------------------------


wt-amt1     ATGGGGAAGTGTTGCAGTTGGCATTCTTGCAAATCCGGAGGTTAACGGATATGCAGGCCT 989
G309S-rev   ATGGGGAAGTGTTGCAGTTGGCATTCTTGCAAATCCGGAGGTTAACGGATATGCAGGCCT 1132
G309S-fwd   ------------------------------------------------------------


wt-amt1     ACTGTTCGGAAATCCGCAACTGCTAGTTTCACAACTGATTGCGGTTGCATCCACAACAGC 1049
G309S-rev   ACTGTTCGGAAATCCGCAACTGCTAGTTTCACAACTGATTGCGGTTGCATCCACAACAGC 1192
G309S-fwd   ------------------------------------------------------------


wt-amt1     CTACGCCTTCCTCGTGACGCTGATACTGGCAAAGGCTGTTGATGCCGCTGTGGGGCTGAG 1109
G309S-rev   CTACGCCTTCCTCGTGACGCTGATACTGGCAAAGGCTGTTGATGCCGCTGTGGGGCTGAG 1252
G309S-fwd   ------------------------------------------------------------


wt-amt1     GGTTAGCTCGCAGGAGGAGTACGTCGGTCTCGACCTGTCGCAGCATGAGGAGGTTGCCTA 1169
G309S-rev   GGTTACCTCGCAGGAGGAGTACGTCGGTCTCGACCTGTCGCAGCATGAGGAGGGGGCCTA 1312
G309S-fwd   ------------------------------------------------------------


wt-amt1     CACGTGA----------------------------------------------------- 1176
G309S-rev   CACGCTCGAG[CACCACCACCACCACCAC]TGAGATCCGGCTGCAACAAACCCGAAAGAAGG 1372
G309S-fwd   ------------------------------------------------------------


wt-amt1     -----
G309S-rev   ATCGG 1377
G309S-fwd   -----
```

8. Outlook

Many aspects of the biological function of Amts are still poorly understood. The goal of this thesis is to provide some evidences to resolve open mechanistic questions. To achieve this point we were doing high-resolution structural and functional studies on wild-type and different protein variants using previously well characterized the Amt-1 from *A. fulgidus* as a model system. These variants should fascilitate analysis of the role of Amt-1 and other Amt proteins in acquisition of ammonium.

For estimation transport activity of Amt-1 we have constructed BL C43(DE3) *E. coli* cells that were knock-out in their *amtB* gene and transformed with plasmid carrying different variant of wild-type *amt-1* from *A. fulgidus*, and developed *in vivo* system for determination ammonium upatke based on the Nessler reaction. The main property of an *ΔamtB* C43(DE3) *E. coli* strains is almost completely lacking of any significant ammonium uptake under our experimental conditions that leads us to conclusion that Amts are indeed necessary for ammonium transport. Our preliminary functional results of different Amt-1 variants demonstrated that they are all capable of ammonium uptake but rate varies among diffent variants.

The interepretation of these results is not straightforward, in particular considering the possibilty of sensitivity of this test system and therefore it is necessary to optimaze this method. There is a lot of work still to do in order to characterize the function Amts. Accordingly; further *in vitro* studies with purified and reconstituted Amt-1 in proteoliposomes will be required, as well. Taken together, *in vitro* and *in vivo* studies, will give a more quantitative understanding of the conduction mechanism in general.

9. Danksagung

This thesis arose in part out three years of research that has been done since I came to Dr. Susana Andrade's group.

In the first place, I owe my deepest gratitude to Dr. Susana Andrade for her interest, advice, constant guidance, motivation and invaluable criticisms from the initial to the final level enabled me to develop an understanding of the subject. Above all and the most needed, she provided me unflinching encouragement and support in various ways. I am indebted to her more than she knows.

My sincere thanks to my co-supervisor Professor Dr. Oliver Einsle for his tutorship on this work, support and ideas. He provided me with direction, technical support. I doubt that I will ever be able to convey my appreciation fully, but I owe him my eternal gratitude

I thank to Professor Dr. Michael Müller from the Institute of Pharmaceutical Sciences, University of Freiburg for accepting the job as co-referee.

Dr. Susana Andrade, Professor Dr. Oliver Einsle and Dr. Stefan Gerhardt deserve one special thank for the moral support and giving me many warmhearted helps in my life

Thanks to all my present and former colleagues for the pleasant working atmosphere at the lab: Camila Hernández Frederick, Claudia Litz, Tobias Pflüger, Tobias Wacker, Anja Pomowski, Bianca Hermann, Dr. Daniel Wohlwend, Juan Du, Julian Seidel, Peer Lukat, Ramona Labatzke, Thomas Spatzal, Wei Lü, Dr. Volodymyr Shnitsar, Dr. Daniel Heitmann, Dr. Maren Hoffmann, Sandra Henning, Elena Kasparyan, Linda Wiliams, Christiane Zähringer-Steffens, Antonio Espin, Helmut Hamacher.

Where would I be without my family? Words fail me to express my appreciation to my Hubert whose dedication, love and persistent confidence in me, has taken the load off my shoulder. I owe him for being unselfishly let his intelligence, passions, and ambitions collide with mine. My parents, Gradimir and Željka are people who put the fundament my learning character, showing me the joy of intellectual pursuit ever since I was a child.

Furthermore, grand-mothers and grand-fathers are the ones who sincerely raised me with their caring and gently love, thank you.

Finally, I would like to thank everybody who was important to the successful realization of thesis, as well as expressing my apology that I could not mention personally one by one.

Danke! Thank you! Хвала!

Daniel Čebo

10. Curricculum Vitae

Biographical information

Name	:	Daniel Čebo
Birth date	:	16. 03. 1978
Nationality	:	Serbian

Education

2007-2010	:	PhD, University of Freiburg, Germany
1998-2003	:	Diploma of Molecular Biology and Physiology University of Belgrade, Serbia

Personal projects

Doctoral thesis: "Site-directed muatagenesis as a tool for functional characterization of the Ammonium Transport Protein Amt-1 from *Archaeoglobus fulgidus*" (at the Institute for Organic Chemistry and Biochemistry, Department of Biochemistry, Freiburg, Germany. September 2010).

Diploma thesis: "Effect of L-Ribavirin on the viability of embryonic and adults astrocytes *in vitro*" (at the Institute for Biological Research ''Siniša Stanković'', Department of Neurobiology and Immunology, Belgrade, Serbia. May 2003).

Honors and Awards

2005-2007	:	Fellowship of the Hertie Foundation
2002-2004	:	Fellowship of Serbian Ministry of Education